T0245162

When vectors collide with cultures: 'anthropo-vector ecology', who is controlling who?

When vectors collide with cultures: 'anthropo-vector ecology', who is controlling who?

– Book of Abstracts –

The 20th European Society for Vector Ecology Conference 2016

–

3-7 October 2016
Lisbon, Portugal

EAN: 978-90-8686-291-7
e-EAN: 978-90-8686-837-7
ISBN: 978-90-8686-291-7
e-ISBN: 978-90-8686-837-7
DOI: 10.3920/978-90-8686-837-7

Cover design: Marcos Santos

First published, 2016

© Wageningen Academic Publishers
The Netherlands, 2016

This work is subject to copyright. All rights are reserved, whether the whole or part of the material is concerned. Nothing from this publication may be translated, reproduced, stored in a computerised system or published in any form or in any manner, including electronic, mechanical, reprographic or photographic, without prior written permission from the publisher:
Wageningen Academic Publishers
P.O. Box 220
6700 AE Wageningen
The Netherlands
www.WageningenAcademic.com
copyright@WageningenAcademic.com

The individual contributions in this publication and any liabilities arising from them remain the responsibility of the authors.

The publisher is not responsible for possible damages, which could be a result of content derived from this publication.

Sponsors

Golden sponsor

Platinum sponsor

Silver sponsors

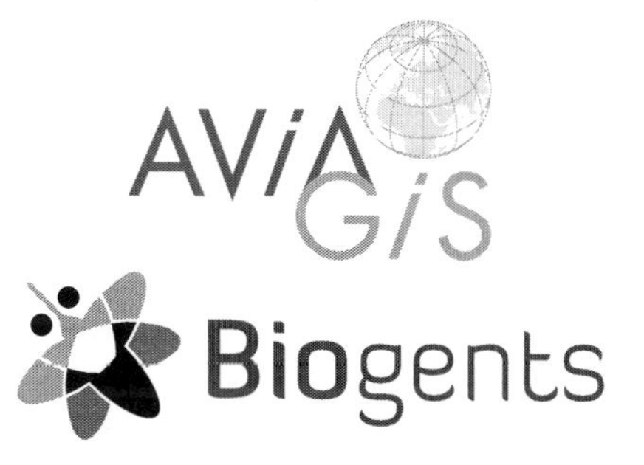

Technical support

How to reach Faculdade de Medicina Veterinaria

By bus

Faculdade de Medicina Veterinaria (FMV) is located in the Polo Universitário do Alto da Ajuda , and you may reach it taking the following buses:

- Bus 723 – Desterro – Algés;
- Bus 729 – Algés – Bairro Padre Cruz;
- Bus 760 – Gomes Freire – Cemitério da Ajuda;
- Bus 742 – B. Madre de Deus – Universidade da Ajuda. Please notice that this bus may end in two different locations (Casalinho Ajuda and Universidade da Ajuda). You must catch the bus to B. Madre de Deus – Universidade da Ajuda.

By underground

- Blue line to Marquês de Pombal Station. Then catch Bus 723 (direction Algés);
- Blue line to Pontinha, Carnide or Colégio Militar Stations. Then catch Bus 729 (direction Algés);
- Red line to São Sebastião Station. Then catch Bus 742 (direction Universidade Ajuda);
- Green line to Martim Moniz, Rossio or Cais do Sodré Stations. Then catch Bus 760 (direction Cemitério da Ajuda).

From the airport

From the airport, you may reach FMV by taking the underground and/or buses:

- Underground: Red line from airport to São Sebastião. Then catch Bus 742 (direction Universidade Ajuda).
- Bus 708: Parque Nações Norte to Martim Moniz. Then catch Bus 760 (direction Cemitério da Ajuda).
- Bus 744: Moscavide to Marquês Pombal. Then catch Bus 723 (direction Algés).
- Bus 783: Portela/Prior Velho to Amoreiras. Then catch Bus 723 (direction Algés).

From Oriente train station

You have two ways of going from Oriente station to FMV:

- By underground (easiest way): Red Line from Oriente to São Sebastião. Then, you must catch Bus 742 to Universidade Ajuda.
  On weekends, you only have Bus 742 available ending on Casalinho Ajuda. Follow the instructions above for this bus.
- By train: catch the train to Alcântara-Terra (this train is available only during the week). Then, you must catch Bus 742 to Universidade Ajuda.

For more information, timetables and itineraries please go to http://metro.transporteslisboa.pt (underground) or http://carris.transporteslisboa.pt (buses).

By taxi

You can always choose to take a taxi from anywhere in the city! These are some phone numbers for taxis:

- Autocoope – (+351) 217 932 756;
- Rádio Táxis de Lisboa – (+351) 219 362 113;
- Teletáxis – (+351) 218 111 100;
- Volancoop – (+351) 218 153 513.

Table of contents

Welcome

When vectors collide with cultures: 'anthropo-vector ecology', who is controlling who?

Dear conference attendees,

On behalf of the European Society for Vector Ecology it is my great pleasure to welcome you all in Lisbon!

This city with its beautiful colours, amazing food, warming people, music, art and culture is going to host our conference for the second time since 1998. I hope you will enjoy your stay and have also the time to get a bit of the vibrant life and energy that is the soul of this incredible city.

The European Region of Society for Vector Ecology has a very important role in getting together scientists form various disciplines with the a common goal of enhancing research, control, dissemination of information and to develop new strategies on vectors and vector-borne diseases. It is very encouraging to see the contributions of well renowned scientists from more than 40 worldwide countries at this conference. I am also very happy to report that our networking of scientists is continually expanding throughout our global community and thus affording all of us new opportunities of collaborations.

These past decades have played an enormous role in discovering new aspects on vector ecology, genetic and genomics of vectors and pathogens they transmit advance technologies on vector control, impact of climate changes, pancontinental vector dispersal and many others. Moreover, the development of new tools for vector surveillance and control is also contributing to the reduction of vector populations and pathogen transmission. The scientific community is also casting light on the importance of defining gaps and needs in medical and veterinary entomology via collaborative projects on a global basis, therefore strengthening our networking activities. But what about the role of social science, anthropology, communication and many other disciplines? How do these factors interact with our fight against vectors? Are vector treatments by professionals in the field more sustainable than citizen participation? Are we using the right language and message to engage communities with science affecting our daily lives?

It was my concern for this year to draw attention to the impact and importance of vectors and vector-borne diseases on landscapes inhabited by human beings and how to make this paradigm most effective to benefit public health. I hope this will help us to re-think about the way we communicate our knowledge, achievements, preventative measures towards vector-borne diseases so that 'non-scientists' will also be able to understand and contribute to their sustainability.

Last but not least, let me express my best appreciation to the local organizing committee for their incredible support and help to make sure that we can deliver a great science during our stay in Lisbon. A very big thanks also goes to all the E-SOVE members and mentors for their continuing support.

Finally, I would like to express my sincere gratitude to all the participants of the 20th E-SOVE conference, without you we could have not make such an outstanding meeting.

Thank you for your continuous support to our society and for the great contributes to successful science, enjoy the conference!

Dr. Eva Veronesi

E-SOVE President
20th E-SOVE Program Chair
The 20th European Society for Vector Ecology Conference 2016

E-SOVE advisory board members

- Bülent Alten, Hacettepe University, Ankara (Turkey)
- Carles Aranda, Baix Llobregat Council, Barcelona (Spain)
- Norbert Becker, KABS, Heidelberg University, Valdsee (Germany)
- Romeo Bellini, CAA, Bologna (Italy)
- Marieta Braks, RIVM, Bilthoven (the Netherlands)
- Major Dhillon, Northwest MVCD, Corona CA (USA)
- Eleonora Flacio, LMA (SUPSI), Bellinzona, (Switzerland)
- Didier Fontenille, IRD, Montpellier (France)
- Jan Lundstrom, Uppsala University, Uppsala (France)
- Douglas Norris, JHBS, Baltimore (USA)
- Francoise Pfirsch, Mosquito Control Organization, Bas-Rhin (France)
- Francis Schaffner, University of Zurich, Zurich and Francis Schaffner Consultancy, Riehen (Switzerland)
- David Sullivan, Zocor Inc, Belgrade (USA)
- Eva Veronesi, University of Zurich, Zurich (Switzerland)
- Marija Zgomba, University of Novi Sad, Novi Sad (Serbia)

Scientific committee

- Fabrizio Balestrino, International Atomic Energy Agency, Vienna (Austria)
- Marieta Braks, RIVM, Bilthoven (the Netherlands)
- Mary Cameron, London School of Hygiene and Tropical Medicine, London (UK)
- Alexandra Chaskopoulou, USDA-ARS, Thessaloniki (Greece)
- Isabel Fonseca, Faculty of Veterinary Medicine, Lisbon (Portugal)
- Claire Garros, Cirad UMR, Réunion Island (France)
- Giovanni Savini, Istituto Zooprofilattico Sperimentale, Teramo (Italy)
- Francis Schaffner, University of Zurich, Zurich and Francis Schaffner Consultancy, Riehen (Switzerland)
- Bill Walton, University of Riverside, California (USA)
- Eva Veronesi, University of Zurich, Zurich (Switzerland)
- Marjia Zgomba, University of Novi Sad, Novi Sad (Serbia)
- Frederic Simard, IRD, Montpellier (France)
- Simon Carpenter, The Pirbright Institute, Pirbright (UK)
- Steve Torr, Liverpool School of Tropical Medicine, Liverpool (UK)

Programme chair

- Eva Veronesi, University of Zurich, Zurich (Switzerland)

Organizing committee

- Isabel Pereira da Fonseca, FMV, Lisbon (Portugal)
- Major Dhillon, NWMVCD, Corona (USA)
- Isabel Lopes de Carvalho, INSA, Lisbon (Portugal)
- Hugo Osório, INSA, Lisbon (Portugal)
- Eva Veronesi, University of Zürich, Zürich (Switzerland)

Local organizing committee

- Isabel Pereira da Fonseca, FMV, Lisbon (Portugal)
- Isabel Lopes de Carvalho, INSA, Lisbon (Portugal)
- Luísa Fernandes, FMV, Lisbon (Portugal)
- Teresa Inácio, FMV, Lisbon (Portugal)
- Hugo Osório, INSA, Lisbon (Portugal)
- David Ramilo, FMV, Lisbon (Portugal)
- Marcos Santos, FMV, Lisbon (Portugal)
- Maria Paula Silva, FMV, Lisbon (Portugal)

Webmaster

- Bill Van Dyke, NWMVCD, Corona (USA)

Scientific programme

Keynote presentations

Theatre

Session 1. Invasive species: ecology and their potential role as vectors

Chairmen: Isabel Carvalho and Hugo Osorio
Date: 3 October – 11.00-13.00

Theatre

Posters

Session 2. Citizen science and social approaches on vector control

Chairmen: Doreen Walther and Aleksandra Cupina
Date: 3 October – 14.30-15.30

Theatre

Posters

Session 3. Epidemiology of vector-borne diseases and their vectors

Chairmen: Isabel Fonseca and Marija Zgomba
Date: 3 October – 16.00-18.00

Theatre

Posters

Session 4. Ecology and behaviour of vectors

Chairmen: Alexandra Chaskopoulou and Francoise Pfirsch
Date: 4 October – 09.00-18.30

Theatre

Posters

Session 5. Emerging vector-borne diseases and risk assessment

Chairmen: Veerle Versteirt and Maria Goffredo
Date: 6 October – 09.00-12.30

Theatre

Posters

Session 6. Vector-pathogens interactions

Chairmen: Gregory Lanzaro and Eva Veronesi
Date: 6 October – 16.00-18.00

Theatre

Posters

Session 7. Taxonomy, systematic and phylogeny of vectors

Claire Garros and Francis Schaffner
Date: 7 October – 09.00-10.30

Theatre

Posters

Session 8. Advance technologies on vector control

Chairmen: Fabrizio Balestrino and Andrea Crisanti
Date: 7 October – 11.00-13.00

Theatre

Posters

Abstracts

Sourcing the crowd for mosquitoes: the use of citizen science for mosquito surveillance

C.J.M. Koenraadt
Wageningen University, Laboratory of Entomology, P.O. Box 16, 6700 AA Wageningen, the Netherlands; sander.koenraadt@wur.nl

Practising science is not an activity that one exclusively performs in a closed off laboratory, looking for wholy grails, silver bullets or 'eureka' breakthroughs. More and more there is a need (or a demand?) to communicate ones research results with the outside world. Indeed, at times it is a daunting task to translate our fundamental results into understandable and tangible pieces for a lay audience. On the other hand, involving the general public may also provide unique opportunities for data collection and result sharing. In this keynote, I will explain how we developed a 'citizen science' approach, called 'Muggenradar' (mosquito radar), to collect data on mosquito nuisance and presence. I will discuss how such 'passive' tools can be complementary to ongoing 'active surveillance' programs, and demonstrate how the collection of large amounts of observations can reveal new insights into mosquito ecology that would otherwise remain obscured. I will also touch on how such approaches can be scaled up and used in different (European) contexts.

The potential of social science research in vector control: an example of women-led tsetse control intervention from northwest Uganda

V. Kovacic[1], S.J. Torr[1], A. Shaw[2], I. Tirados[1], C.T.N. Mangwiro[3], J. Esterhuizen[1] and H. Smith[1]
[1]Liverpool School of Tropical Medicine, Pembroke Place, Liverpool L3 5QA, United Kingdom, [2]Division of Pathway Medicine and Centre for Infectious Diseases, School of Biomedical Sciences, College of Medicine and Veterinary Medicine, The University of Edinburgh, Edinburgh, United Kingdom, [3] Bindura University of Science Education, Department of Animal Science, Bindura, Zimbabwe; vanja.kovacic@lstmed.ac.uk

Community-based interventions, particularly those involving women, have shown a positive impact on the uptake and sustainability of disease control programmes. Those developing and implementing such interventions require skills in qualitative research methods, as well as right skills to approach and work hand-to-hand with the beneficiary communities. There are few examples of how to operationalise such interventions within vector-borne disease control programmes, and we report on a pilot study of a women-led intervention in the context of tsetse control. We developed and implemented a tsetse control intervention in NW Uganda where the control operation was organized and managed by women from beneficiary communities. We evaluated feasibility and sustainability of the intervention using participatory research methods. Our village-based women-led intervention was feasible, cost-effective when compared to an entomology service-managed intervention and showed potential for improved sustainability of tsetse control. Participants demonstrated the motivation, skills and organizational capacity required to manage the control operation. We recommend testing of this operational model within tsetse and other vector borne disease control programmes. We also advocate for more active involvement of social scientists in implementation and evaluation of vector control programmes.

AMSAR: a capacity building project based on the 'training the trainers' concept

C. Silaghi[1], D. Anița[2], J. Bojkovski[3], M. Marinov[4], L. Oşlobanu[2], I. Pavlović[5], G. Savuța[2], P. Simeunović[3], A. Vasić[3] and E. Veronesi[1]
[1]National Centre for Vector Entomology, University of Zurich, Winterthurerstrasse 266a, 8057 Zurich, Switzerland, [2]University of Agricultural Sciences and Veterinary Medicine, Faculty of Veterinary Medicine, 700490 Iasi, Romania, [3]University of Belgrade, Faculty of Veterinary Medicine, Belgrade, Serbia, [4]Danube Delta, National Institute for Research and Development, Tulcea, Romania, [5]Scientific Veterinary Institute of Serbia, Vojvode Toze 14, 11000 Belgrade, Serbia; cornelia.silaghi@uzh.ch

The institutional partnership AMSAR (Arbovirus Monitoring, SurveillAnce and Research; funded by the Swiss National Science Foundation) aims to build up research and monitoring capacities of vectors and vector-borne diseases in Serbia and Romania which faced the emergence of vector-borne pathogens in recent years. The core goals of AMSAR are: (1) to improve research, monitoring and teaching capacities on arthropod vectors and vector-borne pathogens by training young researchers from Serbia and Romania; and (2) to build sustainable scientific connections between Switzerland, Serbia and Romania but also within the countries. AMSAR uses an innovative multiplying approach based on the concept of 'training the trainers'. Two post-doctoral scientists from each partner country received a two-months intensive training in state-of-the-art entomological, molecular and virological techniques at the Swiss partner. Afterwards, these trainers together with the Swiss team and invited experts from the partner countries instruct a larger number of PhD students and young postdoctoral scientists during three application-oriented training schools in Serbia and Romania. Additionally, interdisciplinary skills such as scientific presentation techniques were included. In July 2016, a two-week summer school for 20 participants was implemented in Stara Planina and Belgrade, Serbia, on the subject of morphological vector identification and PCR screening for pathogens. Directly afterwards, a one-week summer school was carried out in the Danube Delta Biosphere Reserve, Romania. Both schools were highly succesfull, with highly motivated students, forming bridges within and between countries. The collected specimens will further be analysed during a third school which will take place in winter 2017 in Iași, Romania.

Bug battles: extending war's metaphors in Sardinia after World War II

M. Hall
University of Zurich, Department of Evolutionary Biology and Environmental Studies, Winterthurerstrasse 190, 8057 Zurich, Switzerland; marc.hall@ieu.uzh.ch

This talk focuses on the Rockefeller Foundation's mid-20th century Sardinian Project, which declared war on *Anopheles* following the invention of powerful insecticides. The goal in Sardinia was not malaria eradication but mosquito eradication, with spray nozzles becoming pesticide guns and airplanes dropping chemical weapons. Although traditionally painted in the colours of glorious victory over human disease, Sardinia's massive eradication program involving tons of pesticides and thousands of health workers also posed drawbacks to humans and their ecosystems. Medicines such as quinine and then atabrine produced negative side-effects, including physiological, economic, and environmental costs. Insecticides such as Paris Green and then DDT created problems for organisms beyond mosquitoes that in turn created problems for people, including the poisoning of fish and birds. This talk explores how the shift from a war against people to a war against insects modified health strategies as well as results. The tentacles of war turned Sardinia's flora and fauna, its mountains and coasts, into an ecological laboratory as well as a political proving ground, changing the face of the earth as well as the structure of the human genome.

First report of *Aedes* (*Stegomyia*) *albopictus* (Diptera: Culicidae) and its establishment in Oran, West of Algeria

K.E. Benallal[1], A. Allal-Ikhlef[2], K. Benhammouda[2], K. Senouci[2], F. Schaffner[3,4] and Z. Harrat[1]
[1]Institute Pasteur of Algeria, Parasitology, Route Petit Staouéli, Dely-Ibrahim, 16047, Algiers, Algeria, [2]Science and Technology University of Oran, Parasitology, El Mnaouar, BP 1505, Bir El Djir 31000, Algeria, [3]Francis Schaffner Consultancy, Riehen, 8006, Switzerland, Switzerland, [4]National Centre for Vector Entomology, Institute of Parasitology, University of Zurich, Rämistrasse 71, 8006 Zürich, Suisse, Switzerland; benallalkamel4@yahoo.fr

Aedes (*Stegomyia*) *albopictus* (Skuse, 1894) has been reported in Tizi Ouzou (East of Algeria) twice in 2011 and 2014. The species' risk map elaborated for ECDC indicated that North Africa is suitable for its establishment. The increase of globalisation of trades, human movements and environmental changes facilitate the introduction and establishment of the invasive Asian tiger mosquito *Ae. albopictus* outside its native geographical area. Alerted by the complaints about mosquito biting which occurred daytime for the inhabitants of the seaside town Ain Turk (West of Algeria), an entomological survey was conducted in December 2015 to determine the origin of this nuisance for which *Ae. albopictus* has been strongly suspected. Thereafter and from April 2016, a monthly survey was performed by regular investigation for eggs and adults using ovitraps and BG sentinel traps at locations where the species has been previously found, in order to confirm whether this species has only been introduced or has established and overwintered. Among mosquitoes trapped in December, specimens of *Ae. albopictus* (2 males, 3 females and 3 pupae) were morphologically and molecularly identified using mitochondrial COI sequences. In May 2016, several eggs were collected by ovitraps at the same place and brought to an insectary and reared to establish a laboratory colony for further analysis. This is the first observation and establishment of the invasive mosquito in the west of Algeria which confirms its presence in the country.

Culicoides imicola: phylogeography, population genetics and invasive status in the Mediterranean basin

S. Jacquet[1], K. Huber[1], C. Chevillon[2], J. Bouyer[1] and C. Garros[3]
[1]Cirad UMR CMAEE, Campus de Baillarguet, 34000 Montpellier, France, [2]CNRS IRD UMR MIVEGEC, Agropolis, 34000 Montpellier, France, [3]Cirad UMR CMAEE, CYROI 2, rue Maxime Rivière, 97490 Sainte-Clotilde, La Réunion, France; garros@cirad.fr

Biological invasions are of major concern because of their environmental, economic and health consequences. Determining and understanding the factors underlying invasion success of species allow predicting potential other biological invasions and developing vector control strategies. *Culicoides* imicola is a major vector species of Orbivirus, including the bluetongue virus (BTV) which affects wild and domestic ruminants. Following BTV emergence in the Mediterranean basin, *C. imicola* populations were recorded in territories where the species was considered to be absent, and consequently was described as expanding its distribution range on a short period. This work aimed at understanding the colonization history of the Mediterranean basin by *C. imicola* and determining the factors underlying the current distribution of the species. The use of a multi-loci approach and the combination population genetics analyses, Approximate Bayesian Computation (ABC) methods and mathematical simulations of the atmospheric dispersion of the species enabled us to: (1) demonstrate that *C. imicola* Mediterranean populations originated from the northern region of sub-Saharan Africa and are established in the Mediterranean basin since the Pleistocene/Holocene period; (2) reveal two sources and routes of colonization in southern Europe, involving the colonization of Iberian Peninsula from Morocco and the incursion in France and Italy from Algeria; and (3) highlight the major role of wind-mediated dispersal and population abundances in the range expansion success of the species. Altogether, our study shed a new light on the timing and routes of colonization of the Mediterranean basin by the bluetongue vector.

Surveillance of invasive mosquitoes at points of entry in Switzerland

P. Müller[1,2,3], L. Vavassori[2,3], L. Engeler[1], T. Suter[2,3,4], E. Flacio[1], V. Guidi[1] and M. Tonolla[1]
[1]University of Applied Sciences and Arts of Southern Switzerland, Applied Microbiology Lab, Via Mirasole 22a, 6501 Bellinzona, Switzerland, [2]University of Basel, Petersplatz 1, P.O. Box, 4003 Basel, Switzerland, [3]Swiss Tropical and Public Health Institute, Vector Group, Epidemiology and Public Health Department, Socinstrasse 57, P.O. Box, 4002 Basel, Switzerland, [4]Avia-GIS, Risschotlei 33, 2980 Zoersel, Belgium; pie.mueller@unibas.ch

Invasive mosquitoes are expanding in Europe. In Switzerland, *Aedes albopictus* was recorded for the first time at the southern tip of Switzerland in 2003, then *Ae. japonicus* in central Switzerland in 2008 and finally *Ae. koreicus* in 2013, again at the south border to Italy. While the Canton of Ticino in the South had a well-established surveillance system for invasive *Aedes* mosquitoes, until 2013 no systematic surveillance was carried out across the rest of the country. This motivated us to initiate a national surveillance programme for invasive *Aedes* mosquitoes and we set up a network with ovi- and BG sentinel traps at 38 potential points of entry, including motorway services stations, ports and airports. The traps were inspected every other week from June to September each year. When positive, eggs or adult specimens were counted and identified to species level by morphology and Matrix Assisted Laser Desorption/Ionization Time-Of-Flight Mass Spectrometry (MALDI-TOF MS). Since the start of the programme, in addition to 8 indigenous species, all 3 invasive *Aedes* species are now also found outside Ticino and it becomes eminent that the European motorway E35, running from Rome, Italy, to Amsterdam, the Netherlands, is a key route of invasion into northern Europe. Worryingly, in 2015 citizens of the Basel area at the border to Germany and France have for the first time submitted specimens that turned out to be *Ae. albopictus*. Equally, the range and numbers of *Ae. japonicus* have increased intensively over the last years and *Ae. koreicus* may now be found across Switzerland – suggesting we haven't seen the end of the invasion yet.

Aedes koreicus: a new European invader and its potential for chikungunya virus

S. Ciocchetta[1,2], N. Prow[3], J. Darbro[2], L. Hugo[2], F. Frentiu[1], F. Montarsi[4], G. Capelli[4], J. Aaskov[1] and G. Devine[2]
[1]Queensland University of Technology, School of Biomedical Sciences and Institute for Health and Biomedical Innovation, 60 Musk Ave, Kelvin Grove, 4059 QLD, Australia, [2]QIMR Berghofer Medical Research Institute, Mosquito control Laboratory, 300 Herston Rd, Herston, 4006 QLD, Australia, [3]QIMR Berghofer Medical Research Institute, Inflammation Biology Group, 300 Herston Rd, Herston, 4006 QLD, Australia, [4]Istituto Zooprofilattico Sperimentale delle Venezie, Diagnostic services, histopathology, parasitology, Viale dell'Università 10, 35020 Legnaro (PD), Italy; silvia.ciocchetta@qimrberghofer.edu.au

Recent outbreaks of arboviral diseases such as chikungunya following the introduction of non-native mosquitoes, demonstrate the public health threat that invasions pose. The discovery of a new but poorly understood mosquito species in Europe, *Aedes koreicus*, demands an investigation of this mosquito's vectorial capacity. During four months of field and lab work conducted in Italy, I compared trapping techniques and used field collected material to establish lab-based colonies of *Ae. koreicus* first in Italy, and then at QIMR Berghofer in Australia. This colony is now used to conduct laboratory experiments on *Ae. koreicus* vector competence for chikungunya virus. The chikungunya (CHIKV) 'La Reunion' strain was administered via artificial oral feeds ($10^{7.5}$ TCID50/ml). Mosquitoes were maintained at two temperatures: 23 °C and a fluctuating temperature simulating a Melbourne summer. Despite the low initial hatching rate of *Ae. koreicus* eggs (10.39±2.41%), the long gonotrophic interval (11.5±4.94 d) and lengthy development times (12.71±0.45 d) our unique colony proved suitable for viral challenges. *Ae. koreicus* were exposed to viremic bloodmeals, dissected (25 mosquitoes on days 3, 10, 14, 21 post-infection) and evaluated for CHIKV infection. Feeding rates (65.5%, N=342) and survivorship (97.3%, N=224) were high. Body and salivary infection rates will be presented. These findings are essential for understanding the public health risks of mosquito invasions, targeting surveillance and control initiatives against invasive mosquito species and understanding the effect of daily temperature fluctuations on vector competence.

Exotic mosquitoes in the Netherlands: control and the impact on society

C.J. Stroo, M. Ter Denge and A. Ibañez Justicia
NVWA, Centre for Monitoring of Vectors, P.O. Box 1902, 6700 HC Wageningen, the Netherlands; c.j.stroo@nvwa.nl

Four species of exotic mosquitoes caught in the Netherlands up to now required control. Although a standard approach is followed in all cases, both the success and the impact of the treatments differ. Early detection as well as species specific biology are key factors that determine the effort needed for eradication. *Aedes atropalpus*, *Ae. aegypti* and also *Ae. albopictus* so far have been stopped before populations established. Their different stories will be told. For *Ae. japonicus*, detection only happened after the species was well established. In 2015 the decision was made to not only eradicate this species in new locations but also to start downsizing the resident population in Lelystad. Targeting hotspots, the many allotment gardens, population reduction is aimed for. Monthly treatment with Vectomax locally lead to strong reductions, as shown by grid based evaluations. Limited reoccurrence in 2016 might merely indicate egg survival. Owners of the allotment gardens are cooperating, making it possible to treat the vast majority of all breeding sites found. Open communication is essential for effective action. Public involvement in removing potential breeding sites is low so far and not asked for outside the hotspots as yet. A small lab evaluation of options that the general public has to treat private rainwater barrels with household products was carried out. The success rate of using household oils, detergent and copper coins was evaluated against the use of Aquatain, since that could theoretically be handed out for doing home treatments. The tests were carried out using *Ae. aegypti* as a model species and yielded negative results. Aquatain was 100% successful in preventing emergence but also resulted in a strong deterring effect on oviposition. Vectomax treatment does not show this effect and treated spots therefor act as a sink for eggs. Combinations of source reduction and larviciding thus are the current approach. The search for the ideal mix of measures continues. Involvement of the public is often needed, therefor careful consideration of how and when to communicate is required. Experiences from the field, both good and troublesome will be discussed.

Accidental introductions of yellow fever mosquitoes (*Aedes aegypti*) to the Netherlands with human trade and transport

A. Ibañez-Justicia[1], W. Den Hartog[1], A. Gloria-Soria[2], J.R. Powell[2], M. Dik[1], F. Jacobs[1] and C.J. Stroo[1]
[1]Centre for Monitoring of Vectors, Food and Consumer Product Safety Authority, P.O. Box 9102, 6700 HC Wageningen, the Netherlands, [2]Yale University, Department of Ecology and Evolutionary Biology, New Haven, CT 06511, USA; a.ibanezjusticia@nvwa.nl

Global spread of yellow fever mosquito (*Aedes aegypti)* has been historically related to human trade and transport. In this way, the species has successfully colonized suitable locations in countries in all continents except the Antarctica. The first record in northern Europe was in the Netherlands in 2010 in two tire yards, but the population was eliminated. Genetic markers associated the introduction with an import of used tires from Miami, Florida (USA). Due the zika outbreak in South America, concern arises about the possibility of new introductions of the mosquito vector into the country. Used tires import is considered the most likely pathway for introduction of yellow fever mosquitoes in the country, via desiccation resistant eggs. Nevertheless, other likely pathways have been analyzed with the aim of promptly detecting accidental introductions in order to react and avoid temporary proliferation, and in the worst case, to avoid transmission of vector-borne diseases. Used tire companies, airports, harbors, greenhouses and flower auctions are routinely surveyed for presence of exotic and invasive mosquitoes. In June 2016, three adult yellow fever mosquitoes were captured in two locations at the airport of Schiphol confirming the possibility of air-borne movement of this mosquito. Following this finding both the number of traps and the monitoring frequency were increased 2-fold. We will present and discuss the results of intensive surveillance following the finding and the control put in place since then, as well as the results from the population genetic analysis of the collected mosquitoes designed to identify the origin of the introduction.

Expansion of the Asian tiger mosquito *Aedes albopictus* in Bulgaria between 2012 and 2015

O. Mikov[1]*, I. Katerinova*[2] *and T. Agoushev*[3]
[1]*National Centre of Infectious and Parasitic Diseases, 26 Yanko Sakazov blvd., 1504 Sofia, Bulgaria,* [2]*National Diagnostic Science-and-Research Veterinary Medical Institute, 15 Pencho Slaveykov blvd., 1000 Sofia, Bulgaria,* [3]*Agricultural University, 12 Mendeleev blvd., 4000 Plovdiv, Bulgaria; mikov@ncipd.org*

The invasive Asian tiger mosquito *Aedes albopictus* was found for the first time in Bulgaria in 2011. The invaded area was the town of Sozopol in Burgas region (South-Eastern Bulgaria), located on the Black Sea coast. Despite the control efforts performed next year, the species was found breeding in the same location in 2012. Here we report the subsequent findings of *Ae. albopictus* during a four-year period following the initial invasion. In 2013, mosquito trapping activities were conducted sporadically in Blagoevgrad and Plovdiv regions using BG-Sentinel traps with BG-Lure. In 2014, the 'National programme for prevention and control of vector-borne infections in humans in the Republic of Bulgaria 2014-2018' has started, establishing a mosquito surveillance network of light traps in 15 regions of the country. Catches were sent for identification to the National Centre of Infectious and Parasitic Diseases (NCIPD) on a monthly basis between April and October. Simultaneously, field collections were conducted by NCIPD in Blagoevgrad, Burgas, Vratsa, Montana, Pleven, Plovdiv, Haskovo and Silistra regions using BG-Sentinel traps and CDC light traps, as well as human landing collections. For the first time *Ae. albopictus* was found outside Burgas region in August 2013. It was trapped by a BG-Sentinel trap in the town of Petrich (Blagoevgrad region, South-Western Bulgaria) close to Greece. In 2014, breeding populations were found in Blagoevgrad, Burgas, Vratsa, Plovdiv and Montana regions after receiving nuisance complaints from citizens. Larvae and adults of *Ae. albopictus* were confirmed in samples from the regions of Burgas, Montana and Stara Zagora. In the regions of Pleven, Haskovo and Silistra it was not found. In all the colonised areas, *Ae. albopictus* was found in settlements with an altitude up to 200 metres. Its establishment in different parts of Bulgaria shows both distribution from one municipality to the neighbouring ones (in Burgas region), and numerous independent introductions from different points of entry.

Update on the invasive mosquito species in Europe, is successful elimination an option?

V. Versteirt[1], W. Tack[1], F. Schaffner[1], H. Kampen[2], D. Petric[3], V. Robert[4] and G. Hendrickx[1]
[1]Avia-GIS, Risschotlei 33, 2980 Zoersel, Belgium, [2]The Friedrich-Loeffler-Institut, Südufer 10, 17493 Greifswald, Germany, [3]Novi Sad University, Dr Zorana Đinđića 1, 21000 Novi Sad, Serbia, [4]Institut de recherche pour le développement, Boulevard de la Lironde 2214, 34394 Montpellier, France; vversteirt@avia-gis.com

Over the last decades, new challenges have emerged for national public health authorities across Europe due to both the introduction of highly invasive vector species as the number of reported vector-borne disease cases increased severely. Tropical diseases transmitted by invasive *Aedes*-mosquitoes such as chikungunya (CHIK) fever, dengue and recently also zika are expanding worldwide. It is therefore crucial that the status of these invasive mosquitoes are monitored, underlining the importance of Pan-European projects such as VectorNet and its preceding project VBORNET. During the summer of 2015, 10 field projects were initiated, driven by a spatial gap-analysis highlighting knowledge gaps in the invasive species distribution. It was the first simultaneous field activity across Europe involving several teams and led by the VectorNet consortium. Standardised methods specifically designed for these studies and based on the ECDC guidelines were used. As a results, not only local capacity has been built but the status of the invasion of several important *Aedes* species could be given. It is not only important to get the distribution of invasive species right, it is also important to be able to act in time. The latter seems to be the most difficult part of studying vectors and once an invasive mosquito has established it is often impossible to eliminate it. We hereby present an example of a successful control campaign against *Aedes japonicus* in Belgium. These activities could only be possible due to recurrent political questions and the information available via VBORNET/VectorNet.

Predicting the spatial-temporal risk of Asian tiger mosquito (*Aedes albopictus*) introduction into Germany

R. Lühken[1], A. Jöst[2], I. Schleip[3], U. Obermayr[3], A. Rose[3], N. Becker[2,4] and E. Tannich[1,5]
[1]Bernhard Nocht Institute for Tropical Medicine, Bernhard-Nocht-Straße 74, 20359 Hamburg, Germany, [2]German Mosquito Control Association (KABS), Institute for Dipterology, Georg-Peter-Süß-Str. 3, 67346 Speyer, Germany, [3]Biogents AG, Weißenburgstr. 22, 93055 Regensburg, Germany, [4]University of Heidelberg, Grabengasse 1, 69117 Heidelberg, Germany, [5]German Centre for Infection Research (DZIF), partner site Hamburg-Luebeck-Borstel, Inhoffenstraße 7, 38124 Braunschweig, Germany; renkeluhken@gmail.com

In the course of continuing eco-climatic change and globalization, invasive mosquito species (IMS) are of growing importance in many countries of Europe. Over the last decade, three IMS were also detected in Germany: *Aedes albopictus* (in the year 2007), *Ochlerotatus japonicus* (2008), and *Ae. koreicus* (2015). A systematic evaluation of service stations along different main motorways in Southern Germany indicated regular and repeated introduction of *Ae. albopictus* into Germany. The specimens probably originate from established source populations in southern Europe and are introduced by transit road traffic. Thus, the risk of introduction should be basically influenced by two factors: (1) the density of *Ae. albopictus* in southern Europe; and (2) the amount of traffic. However, the relative importance of these factors is unclear. Therefore, in this study it was analysed, whether the spatial-temporal detection of *Ae. albopictus* specimens at the service stations in Southern Germany over the last three years could be explained by statistical models including the phenology of the source population in southern Europe and the seasonal pattern of the traffic volume. The analysis clearly demonstrated that the phenology of the source population is probably the most important factor driving the introduction risk of *Ae. albopictus* into Germany. In combination with risk maps for the establishment in Germany, this model can be used to identify time periods and areas, which should be in the focus of surveillance programs for the Asian tiger mosquito in Germany.

Exotic invasive mosquitoes: introduction pathways into Switzerland

E. Flacio[1], L. Engeler[1], T. Suter[2], P. Müller[1,2,3] and M. Tonolla[1]
[1]University of Applied Sciences and Arts of Southern Switzerland, Laboratory of Applied Microbiology, via Mirasole 22 a, 6500 Bellinzona, Switzerland, [2]Swiss Tropical and Public Health Institute, Epidemiology and Public Health Department, Socinstrasse 57, 4002 Basel, Switzerland, [3]University of Basel, Petersplatz 1, 4003 Basel, Switzerland; eleonora.flacio@supsi.ch

The extensive surveillance system on *Aedes albopictus* (*Stegomyia albopicta)* ongoing in Canton Ticino (Southern Switzerland) since the year 2000, permitted to have a close follow up of introduction and dispersal pathways into the territory of this exotic invasive mosquito species. Differences in introduction are related to traffic pressure, whereas variation in establishment to climatic conditions and availability of potential breeding sites. Other container breeding mosquitoes, as *Ae. koreicus* and *Ae. japonicus*, are detected since 2013 in parallel with *Ae. albopictus* and seem to have similar introduction pathways. Understanding the invasive mosquitoes spread permits to assess priorities both in surveillance and control measures strategies.

Actual and potential distribution of two invasive mosquito species in Slovenia

K. Kalan, E. Bužan and V. Ivović
University of PrimorskaFaculty of Mathematics, Natural Sciences and Information Technologies, Department of Biodiversity, Glagoljaška 8, 6000 Koper, Slovenia; katja.kalan@upr.si

Apart from the most dispersed, *Aedes albopictus*, there are five alien aedine mosquito species spreading in Europe. They represent a considerable threat on public health as they can serve as vectors of exotic pathogens. In Slovenia we performed a survey in 2013 and 2015 to search for invasive mosquito larvae in artificial water containers at the cemeteries, around human dwellings and in used tires at vulcanizing companies. Additionally, passive reports from citizens were included. With the study we have confirmed the presence of two invasive mosquito species in Slovenia, *Ae. albopictus* and *Ae. japonicus*. In 2013, *Ae. albopictus* was mostly present in Southwestern and central part of the country, with some isolated locations in other parts of Slovenia. Nevertheless, *Ae. japonicus* was found in a large part of the investigated area. This species was previously collected from a small area near the Austrian border in 2011, and in a few years it has colonized the majority of Northeastern part of Slovenia. We did modelling of their potential future distribution of both species. The additional mosquito's distribution survey was repeated in 2015, where the results of the models from 2013 were verified. The distribution of *Ae. albopictus* was consistent with the model's results, but *Ae. japonicus* colonized a far wider area that the model predicted and spread significantly in the country from 2013 to 2015. Still, we do not have systematic mosquito surveillance or control in coutry. *Ae. albopictus* and *Ae. japonicus* are spreading fast and they both present high nuisance for the humans, as well as a public health risk since transmitting vector borne diseases. Therefore an effective national monitoring program of invasive mosquito species is highly needed. With well planned, long term strategy we will be able to detect and control their occurrence, monitor their pathogens and prevent the establishment of new foci of mosquito-borne diseases.

Invasive mosquito species love Italy: a history of invasion

F. Montarsi[1], S. Ravagnan[1], A. Drago[2], S. Martini[2], D. Arnoldi[3], F. Baldacchino[3], A. Rizzoli[3] and G. Capelli[1]
[1]Istituto Zooprofilattico Sperimentale delle Venezie, Parasitology department, Viale dell'Università, 10, 35020 Legnaro, Padua, Italy, [2]Entostudio srl, Viale Del Lavoro, 66, 35020 Ponte San Nicolò, Padua, Italy, [3]Fondazione Edmund Mach, Centro Ricerca e Innovazione, Via E. Mach, 1, 38010 S. Michele all'Adige (TN), Italy; gcapelli@izsvenezie.it

Invasive mosquito species (IMS) belonging to genus *Aedes* are repeatedly recorded out of their native places due to global trade and human travel. They are able to survive during long-distance transport as eggs and can adapt to a new environment. Invasive *Aedes* species are potential vectors of arboviruses and their establishment in new areas pose a threat for human and animal health. Several invasive *Aedes* species are now established in Europe and Italy is one of the most infested countries. *Aedes albopictus*, represents the best example of success of colonization in Europe. It arrived for the first time in Albania in 1979 but it found the best conditions in Italy, spreading in whole country after its first establishement in 1991. *Aedes koreicus*, another IMS was found in 2011 in north-eastern Italy. It is spreading from the original infested area towards South and West and is now present in four Italian Regions. Lastly, a third IMS was detected in 2015, again in north-eastern Italy: *Aedes japonicus japonicus*, found in July 2015 in three sites in Udine province and confirmed in March 2016. These three species develop in the same breeding sites and are often found in artificial containers together with other mosquito species. In the same area other invasive species, *Ae. atropalpus* (1996) and *Ae. aegypti* (1972) were recorded but they did not establish thanks to mosquito control measures. It is interesting to note that Veneto is the region with the most frequent experience of invasive mosquito introduction in Italy, likely as a consequence of the intensive trade of goods and thanks to regular mosquito surveillance system. The establishment of new invasive vectors of diseases complicates the current surveillance and requires well trained personnel for identification.

First record and establishment of the mosquitoes *Aedes albopictus* and *Aedes japonicus* in Strasbourg (North-East of France) in 2015

C. Bender[1], O. Pompier[1], F. Pfirsch[1], E. Candolfi[2] and B. Mathieu[2]
[1]SLM67, Syndicat de Lutte contre les Moustiques du Bas-Rhin, Institut de Parasitologie, 3 rue Koeberle, 67000 Strasbourg, France, [2]Institut de Parasitologie et de Pathologie Tropicale, Faculte de Médecine de Strasbourg, Institut de Parasitologie, 3 rue Koeberle, 67000 Strasbourg, France; fpfirsch@demoustication-bas-rhin.fr

The first established population of *Aedes albopictus*, the Asian tiger mosquito, vector of dengue and chikungunya, was recorded in southern France in 2005. Since 2006, a coordinated surveillance program has been established at national level, combining epidemiological surveillance, entomological survey of invasive species and vector control measures. In the region of Strasbourg (48°35'N, 7°44'E, altitude 142 m), the SLM67 (local mosquito control organisation) has been in charge of entomological survey since 2010, and cooperation with the University of Strasbourg began in 2013. The first detected species was *Aedes japonicus* in September 2014, which was followed by an isolated detection of *Aedes albopictus* in October. Reinforced monitoring using ovitraps and citizen participation was organised in 2015. Control measures failed to eradicate the species, which have been declared present since 2015. In-progress surveillance and control programs are described.

First record of *Aedes albopictus* (Skuse, 1894) in Lampedusa, Linosa and Pantelleria islands and current distribution in Sicily, Italy

R. Romi[1], S. D'Avola[2], D. Todaro[2], L. Toma[1], F. Severini[1], A. Stancanelli[3], F. Antoci[3], F. La Russa[3], D. Boccolini[1], S. Casano[4], E. Carraffa[5], F. Schaffner[6,7], M. Di Luca[1] and A. Torina[3]
[1]Istituto Superiore di Sanità, Dept. of Infectious, Parasitic and Immunomediated Diseases, Viale Regina Elena, 299, 00161, Rome, Italy, [2]ASP Trapani, Dept. of Veterinary Prevention, c.da Kamma 91017, Pantelleria, Italy, [3]Istituto Zooprofilattico Sperimentale della Sicilia, Via Gino Marinuzzi 3, 90129 Palermo, Italy, [4]City Council of Pantelleria, P.zza Cavour, 91017, Pantelleria, Italy, [5]City Council of Lampedusa and Linosa, Via Cameroni, s.n.c, 92010 Lampedusa e Linosa, Italy, [6]University of Zurich, Institute of Parasitology, Winterthurerstrasse 266 a, 8057 Zurich, Switzerland, [7]3Avia-GIS, Risschotlei 33, 2980 Zoersel, Belgium; alessandra.torina@izssicilia.it

In order to verify the current southern limit of *Aedes albopictus* distribution in Europe, an entomological study was carried out within the VectorNet Project, founded by Avia-GIS, by the Istituto Superiore di Sanità (Rome) and the Istituto Zooprofilattico Sperimentale della Sicilia (Palermo). The study was carried out in Agrigento, Siracusa, Ragusa, Trapani and Caltanissetta provinces from June to October 2015. Sporadic surveys were carried out in July and in October in Lampedusa, Linosa and Pantelleria islands. Human bait collections were performed and CDC traps, BG-Sentinel® traps and ovitraps were used. In Lampedusa, *Ae. albopictus* was found using 12 ovitraps in 9 selected sites managed for nearly two weeks in July: all sites were positive. In October, 5 ovitraps in 5 sites were monitored for 3 days and only one resulted positive. In Linosa, during the survey in October, the species was found in the village and in the nearby cemetery. In Pantelleria, *Ae. albopictus* was collected in 8 sites, monitoring 20 sites with 22 ovitraps during two surveys in October. *Ae. albopictus* was also recorded in all the investigated Sicilian provinces. Present study reports for the first time *Ae. albopictus* presence in Pantelleria, Linosa and Lampedusa, which represent the Southernmost European limit of this species and its occurrence in the whole Sicily. A monitoring activity during the whole year and a surveillance in ports and airports should be implemented.

Tolerance of *Aedes albopictus* and *Ae. aegypti* eggs to different concentrations of water salinity

R. Lühken[1], A. Plenge-Bönig[2] and E. Tannich[1,3]
[1]Bernhard Nocht Institute for Tropical Medicine, Bernhard-Nocht-Straße 74, 20359 Hamburg, Germany, [2]Institute for Hygiene and Environment, Marckmannstraße 129a, 20539 Hamburg, Germany, [3]German Centre for Infection Research (DZIF), partner site Hamburg-Luebeck-Borstel, Inhoffenstraße 7, 38124 Braunschweig, Germany; renkeluhken@gmail.com

The global spread of emerging mosquito-borne pathogens (e.g. zika virus), which are predominantly transmitted by different invasive mosquito species, indicates the demand for a complete risk assessment of the man-made introduction of these vector species. Tires as well as other small water containers (e.g. for plants) are known to facilitate the spread of invasive mosquitoes all over the world. However, also on the vessels themselves numerous small water bodies containing rain, seawater from splashes or both can be found, e.g. small cavities on the tarpaulins of rescue boats or on containers. These probably hold water long enough to serve as potential breeding sites for mosquito species. However, it is not known whether these containers with at least slightly brackish water allow the survival of *Aedes albopictus* and *Ae. aegypti* eggs. Therefore, the hatching success of eggs from both species experimentally exposed to water with different salinities (from tap water up to sea water) were compared. Our study showed that the survival rates significantly decrease with increasing salinities, but that slightly brackish water allow a complete survival to adults. In conclusion, besides the goods transported, the vessels themselves have to be considered to provide suitable breeding sites allowing the worldwide spread of invasive mosquito species.

Current situation of the *Aedes aegypti* and *Aedes albopictus* invasion of Eastern Blacksea area in Turkey

M.M. Akiner[1], M. Ozturk[1], B. Demirci[2], V. Robert[3] and F. Schaffner[4]
[1]Recep Tayyip Erdoğan University Faculty of Arts and Sciences, Biology, Recep Tayyip Erdoğan University Zihni Derin Campus Fener, 53100, Turkey, [2]Kars Kafkas University Faculty of Arts and Sciences, Molecular Biology, Kars Kafkas University Faculty of Arts and Sciences Kars, 36000, Turkey, [3]IRD, IRD Montpellier, 1, France, [4]Avia GIS, Risschotlei 33, 2980 Zoersel, Belgium; akiner.m@gmail.com

The region of Eastern Blacksea stated eastern border of Turkey with Georgia and Armenia. This area is an important for agriculture and tourism and trade between Europe to Caucasian area. Invasion survey in 3 cities was conducted in september 2015 and from april to july 2016. 126 locations were surveyed in 2015 september. Larval collection was performed inside used tyre and adult collection was performed human bait and EVS trap. *Ae. albopictus* and *Ae. aegypti* were found 12 locations and also found together three locations. We checked again the same areas after winter season. Investigation results in 2016 showed overwintering succcess for *Ae. albopictus* and *Ae. aegypti* for the areas. After winter season location characteristics were changed and tyres transferred to another areas or sold. Approximately 40 km from Georgian border to west areas are including *Ae. albopictus* and spreading new points. *Ae. aegypti* were found two cities five points in 2016 and couldnt find yet any sample for 2015 collection points from border zone (40 km from border) in 2016. Three points of Pazar city including pure *Ae. aegypti* and two points of Trabzon city (new points of *Ae. aegypti)* including mix populations of *Ae. albopictus.* From 2015 to 2016 *Ae. albopictus* spreading whole areas from 40 km Georgian border and found generally inside used tyre. We also found new points west side of the Trabzon after last collection points in 2015 located Giresun city border. *Ae. ageypti* was found in new points in Trabzon city from used tyre storage areas 120 km away last collection points in 2015. This study is supported by Vectornet.

Mobocon: program for strengthening sustainable mosquito-borne disease control in Dutch Caribbean

M. Braks[1], K. Hulshof[2], T. Leslie[3], M. Henry[4] and Y. Gerstenbluth[5]
[1]National Institute of Public Health and the Environment, Center Infectious Disease Control, A. van Leeuwenhoeklaan 9, 3720 BA Bilthoven, the Netherlands, [2]Island Government of Saba, Department of Public Health, No.1 Power Street, The Bottom, Saba, Netherlands Antilles, [3]Island Government of St. Eustatius, Public Health Department, Cottage Road, Oranjestad, Sint Eustatius, Netherlands Antilles, [4]Ministry of Public Health, Social Development and Labor, Hygiene and Vector Control, P.O. Box 943, Philipsburg, Sint Maarten, Netherlands Antilles, [5]Ministry of Health of Curacao, Communicable Diseases Unit, Piscaderaweg 49, Willemstad, Curacao, Netherlands Antilles; marieta.braks@rivm.nl

The current spread of zika virus infections in the Americas is the latest illustration of the re-emergence of mosquito borne diseases. Driven by trade and travel on worldwide scale, climate change, and continuous urbanisation, the world has been witnessing an increase in diseases that are mainly transmitted by the yellow fever mosquito, *Aedes aegypti*. The response of the individual Dutch Caribbean islands to zika virus outbreaks was similar to those of chikungunya and dengue in preceding years. Responses include increased disease surveillance, intensified effort on source reduction in public and private spaces, outbreak communication, and intersectorial collaboration (port, hotels, cruises). The level of the response depends on the ecological, political, cultural and vectorial situation of individual islands that differ largely. As a consequence, the tools, support and the resources of the islands devoted to vector control also differ enormously. During the last three years, the contact between the Caribbean and European Netherlands with, respect to sustainable integrated vector control (IVM), has been intensified due to the: (1) implementation of the International Health Regulations (IHR); (2) CHIKV outbreak of 2013-2014; (3) exchange of students; (4) current ongoing zika outbreak in the Americas; and (5) renewed Pan European interest in vector control due to the zika treat. Here we present a joint initiative to strengthening sustainable mosquito-borne disease control in Dutch Caribbean (Mobocon) and to contribute to the region as a network to ultimately be able to face the enormous challenge to control of *Aedes*-borne diseases.

Anopheles plumbeus (Stephens, 1828) in Germany: occurrence and public awareness of a nuisance and potential vector species

E.C. Heym[1], H. Kampen[2] and D. Walther[1]
[1]Leibniz Centre for Agricultural Research, Eberswalder Str. 84, 15374 Muencheberg, Germany, [2]Friedrich-Loeffler-Institut, Südufer 10, 17493 Greifswald-Insel Riems, Germany; eva.heym@zalf.de

Although data are scarce, *Anopheles plumbeus*, a potential vector of malaria parasites and West Nile virus, is considered widely distributed throughout Europe. Due to using tree-holes as natural breeding sites, it generally occurs at low population densities. However, man-made environmental changes providing spacious artificial larval habitats, such as unattended manure pits, have given rise to increasingly frequent incidents of mass development and contributed to the species becoming a serious pest nuisance in certain areas of Europe. Based on active and passive monitoring, we updated the distribution of *An. plumbeus* in Germany and assessed recent incidents of mass development and nuisance. For the latter, information by participants to a German citizen science project, the 'Mueckenatlas', has been analysed. Submitters of *An. plumbeus* were addressed by a questionnaire in order to identify factors describing the probability of an *An. plumbeus* outbreak. In addition, general knowledge about sources and management of mass development was enquired. Our results show that the observation of diurnally active black mosquitoes in connection with a close-by disused farm is well suited to identify mass occurrence of *An. plumbeus*. Often, the study participants were characterised by limited awareness of sources of mosquito development in their neighbourhoods, with particularly little knowledge about the existence of artificial breeding sites. Thus, more public education on mosquito ecology would probably help to prevent and control public health problems linked to *An. plumbeus*.

Citizen science against globalized mosquito-borne diseases

J.R.B. Palmer[1,2], A. Oltra[1,3], F. Collantes[4], J. Delgado[4], J. Lucientes[5], S. Delacour[5], M. Bengoa[6], R. Eritja[1], M.A. Community[7] and F. Bartumeus[1,3,8]
[1]CREAF, Campus UAB, 08193 Cerdanyola del Valles, Spain, [2]Pompeu Fabra University, Ramon Trias Fargas 25-27, 08005 Barcelona, Spain, [3]Centre d'Estudis Avançats de Blanes (CEAB-CSIC), Spain, Acces Cala Sant Francesc 14, 17300 Blanes, Spain, [4]Universidad de Murcia, Campus de Espinardo, 30100 Murcia, Spain, [5]Universidad de Zaragoza, Calle de Pedro Cerbuna, 12, 50009 Zaragoza, Spain, [6]Universitat Illes Balears, Carr. de Valldemossa, km 7.5, 07122 Palma, Spain, [7]Anonymous citizen scientists collaborating through the Mosquito Alert platform, Mosquito Alert, Platform, Spain, [8]Institució Catalana de Recerca i Estudis Avançats, Passeig de Lluís Companys, 23, 08010 Barcelona, Spain; fbartu@ceab.csic.es

We present a scalable citizen science system with the potential to revolutionize the research, surveillance and management of mosquito-borne diseases by adapting to the complex, human-facilitated spreading processes that characterize many invasions. We show the value of citizen science, focusing on the tiger mosquito in Spain, a potential vector of zika, dengue, and chikungunya. In particular, we demonstrate how citizen science has helped to overcome three key challenges posed by the tiger mosquito invasion: (1) early warning as the invasion front spreads to new areas; (2) measuring population distribution in terms of human-vector encounter probabilities; and (3) investigating the human role in dispersal. After only two years of operation, Mosquito Alert has triggered public health and mosquito control protocols in about 174 municipalities. Mosquito Alert estimates of the tiger mosquito population distribution in Spain are comparable to those obtained from traditional surveillance methods, but the Mosquito Alert estimates cover a vastly larger territory and in a scalable manner. The broad geographic coverage of the Mosquito Alert estimates, combined with information on inter-province commuting, allows us to map, for the first time ever, the sources and sinks of tiger mosquito flows in Spain.

Door to door project, a network against tiger mosquitoes

G.H. Karabas[1], C. Matrangolo[1], A. Maffi[2], C. Venturelli[2] and F.N. Bienvenue[3]
[1]CAA 'G. Nicoli', Via Argini Nord, 40014 Crevalcore (BO), Italy, [2]Local Health Unit of Romagna, Via M. Moretti, 47521 Cesena (FC), Italy, [3]Foreigners Center, Via Dandini, 47521 Cesena, Italy; claudio.venturelli@auslromagna.it

The door to door project 'A network against tiger mosquitoes' is an experimental project created by the co-organization of Environment Bureau, Social Services and Center for Foreigners of the Municipality of Cesena; the Department of Public Health of Ausl of Romagna and Fiorenzuola Neighborhood of Cesena. The coordination group includes an entomologist, an operator with a psychology degree, two biologists and a mediator from Cameroon. Since January 2016 Ausl of Romagna, in the frame of Life CONOPS project activities, began the project planning by figuring out the best method for reaching citizens and help them fight mosquitoes. Infact, the main goal of this project is teaching everyone simple methods for eliminating these insects so that in addition to prevention measures taken by the Municipality and Public Health Unit in public areas, citizens manage to take the right measures in their private areas where public units are not authorized to intervene (except particular cases). Since mosquitoes are a public problem which concerns everybody the project involves volunteers with various backgrounds and adopt the peer to to peer education method. The volunteers were chosen by Social Services and Center for Foreigners – SPRAR project (protection service for refugees and asylum seekers) of the Municipality of Cesena. There are ten active volunteers: five from Cesena and five with different origins like Nigeria, Mali and Pakistan. They attended an intensive formation course of five days in April, with additional days organized for the foreigners and montly meetings for updates for all of them. They work in groups of two or three; in each group there is at least one italian and one foreigner. They visit citizens residing in private areas and give them information about prevention methods and easy tricks to apply to avoid the development of larval breeding sites. The volunteers have started going on field in May and will continue to work until the end of September. Each week, they turn in observation sheets to coordination group for data elaboration as well as meeting them for feedback every two weeks.

Contribution of citizen involvement in the surveillance of *Aedes albopictus* in metropolitan France

Y. Perrin[1], F. Jourdain[1], A. Godal[2], G. Besnard[3], P. Bindler[4], R. Foussadier[3], K. Grucker[1], C. Jeannin[5], F. Pfirsch[6], C. Tizon[5] and G. L'Ambert[5]
[1]Centre national d'Expertise sur les Vecteurs, 911, avenue Agropolis, 34394 Montpellier, France, [2]French Ministry of Social Affairs and Health, 14, avenue Duquesne, 75007 Paris, France, [3]EID Rhône-Alpes, 31 Chemin des Prés de la Tour, 73310 Chindrieux, France, [4]Brigade Verte du Haut-Rhin, 92, rue du Maréchal de Lattre de Tassigny, 68360 Soultz, France, [5]EID Méditerranée, 165, avenue Paul Rimbaud, 34184 Montpellier, France, [6]Syndicat Mixte de Lutte contre les Moustiques du Bas-Rhin, 3, rue Koeberlé, 67000 Strasbourg, France; yvon.perrin@ird.fr

The surveillance of *Aedes albopictus*, implemented in France since 1998, has initially included the surveillance of the used tires importers and a network of ovitraps along the major highways and the Italian border. Since its first establishment in 2004, the distribution area has been growing continuously. During all these years, more and more observations have been made as a result of citizen reports. In order to promote the citizen involvement, a web portal has been set up in 2014. This website allows the population to get information about the biology of the species, and to formalize their reporting. The process begins with 3 questions about the size, the color and the presence of a proboscis on the specimen, in order to filter an important part of misleading reports. Information on the place of observation and a photo or a specimen are requested. The identification of the species from a city where it is not known as established leads to entomological investigation in order to evaluate the importance of the colonization, with the view to controlling the species if possible. In 2014 and 2015, 2,276 notifications were received on the website. Among these, 422 were identified as *Ae. albopictus*, concerning 254 newly colonized cities, and 10 new departments. It also contributed to the detection of *Ae. japonicus* in several cities. The citizen involvement provides key information for the surveillance of *Ae. albopictus*. This kind of reporting has become the main detection tool of the species at a distance from the colonized area (>50 km). This method is also particularly sensitive in densely populated areas where ovitraps are in competition with many breeding sites.

Interactive identification key for female mosquitoes (Diptera: Culicidae) of Euro-Mediterranean and Black Sea Regions

F. Gunay[1,2], M. Picard[1] and V. Robert[1]
[1]IRD, MIVEGEC unit (Infectious Diseases and Vectors: Ecology, Genetics, Evolution and Control), UMR 5290 CNRS-IRD-UM & UR224 IRD, 911 Avenue Agropolis, BP 64501, 34394 Montpellier cedex 5, France, [2]Hacettepe University, Biology Department, Hacettepe University, Biology Department, Vector Ecology Research Group, 06800, Beytepe, Ankara, Turkey; gunayf@gmail.com

In the context of the recent introduction of invasive species, an urgent need for a user-friendly identification key for mosquitoes has resurfaced among entomologists and health care officials. As the MediLabSecure medical entomology network team, our experience has shown that it is essential to have a tool that can be used freely, online and offline, by both specialists and non-specialists. 134 mosquito species are included in this tool which are distributed in 65 countries in Europe, North Africa, the Black Sea area and Middle East. Morphological descriptors for female identification are used to build a comprehensive database. This morphological database was consequently edited using the computer software Xper[2], a useful tool to create interactive and easy to use identification keys. This interactive morphological identification key was developed using 57 morphological characters (descriptors). Geographic distribution is another very useful character where more than one country could be selected. All descriptors have detailed description, along with a picture or an illustration. Priority was to line up the descriptors, considering non-experts will use the tool so characters that are more visible than others come first. It allows the user to see the descriptors, states, definitions, images and all taxa on the same page. Training sessions organized in the frame of the MediLabSecure Project, with participants from many of the selected countries in this task, have brought to light that a tool to train inexperienced users is a must. This identification key will be available in January 2017. By giving the possibility to work with a subset of body parts, this key presents the added value of working even when some body parts are missing on the sample. The tool will be upgraded to include a key for fourth instar larva. Moreover, entomologists with variable degrees of experience will validate the key.

Effect of altitude on Simulidae species richness in rivers of Valencian autonomous region

D. López-Peña, J. Herrezuelo-Antolín and R. Jiménez-Peydró
Institut Cavanilles (Universitat de València), Laboratorio de Entomología y Control de Plagas, C/Catedrático José Beltrán Martínez, 2, 46980 Paterna (Valencia), Spain; david.lopez@uv.es

The family Simuliidae, is a group of Diptera which shows health and veterinary importance because of there are certain species which display hematofagic behavior that affects both livestock and humans. This reason added to the lacking knowledge of the family in the Valencian community, have promoted this study in order to provide new bioecology data of this group in these latitudes. The main aim is to demonstrate whether the different altitudes of a river can by some way affect the composition of the blackflies community in that river. Or, on the contrary, there is no relationship between the two variables that may explain the difference of species richness between diferent altitudinal profiles of a river. However, the results show that there does exist a clear correlation between species richness and the altitude variable in several rivers studied of the Valencian Comunity.

Control and monitoring of mosquitoes in Burriana (Castellón, Spain)

A. Lis-Cantín, J. Herrezuelo-Antolín, D. López-Peña and R. Jiménez-Peydró
Institut Cavanilles (Universitat de València), Laboratorio de Entomología y Control de Plagas, C/Catedrático José Beltrán Martínez, 2, 46980 Paterna (Valencia), Spain; alvaroliscantin@gmail.com

Monitoring and control of mosquito populations in urban environments is one of the priorities of action against the risks that it can produce in human population. This fact is compounded in those municipalities where the establishment of *Ae. albopictus* can lead to significant risks. Establish the efficacy of treatments depending on the species phenology that manifest themselves, in each locality, can carry out effective control programs that maintain population levels acceptable. In this study, it is showed data monitoring mosquito populations in a municipality, with presence of *Ae. albopictus*, and the effectiveness of the treatments carried out by third generation biocides are offered, as well as analysis frequencies and efficiencies obtained in the 2016 campaign.

Surveillance of natives and invasive mosquitoes in the Castellón province (Spain)

J. Herrezuelo-Antolín, A. Lis-Cantín, D. López-Peña and R. Jiménez-Peydró
Institut Cavanilles (Universitat de València), Laboratorio de Entomología y Control de Plagas, C/Catedrático José Beltrán Martínez, 2, 46980 Paterna (Valencia), Spain; jaime.herrezuelo@gmail.com

Since the introduction of the tiger mosquito (*Aedes albopictus*) in Valencian Autonomous region in 2009 in the municipality of Torrevieja (Alicante, Spain), the species has had a rapid expansion throughout its territory and in the last two years the colonization of the territory has been significantly increased. Consequence of the expansion and the many problems caused in 2015, the Autonomous Goberment decided to establish an action plan linked to the monitoring and control of *Ae. albopictus*, as well as other species whose possibility of disease transmission are present in the territory. In the present study the results of the first year of the campaign by making an executable plan developed for implementation over a long period of time are offered.

Effect of diflubenzuron mosquitoes metamorphosis

R. Jiménez-Peydró, C. Alamilla-Andreu, S. Torralba-Pertinez and D. López-Peña
Institut Cavanilles (Universitat de València), Laboratorio de Entomología y Control de Plagas, C/Catedrático José Beltrán Martínez, 2, 46980 Paterna (Valencia), Spain; jimenezp@uv.es

Insect growth regulators as third generation insecticides have emerged as a new weapon for insect control. They have been developed as a result of rational leads from basic biochemical research of metabolic disruptors, moult inhibitors and behavioural modifiers of insects. In this group, diflubenzuron, a chitin synthesis inhibitor, has shown great efficacy for mosquito control in breaking the life cycle of the insect leading to the death of larvae and pupae. In the present work there offer the results obtained after the continued studio of three populations submitted, each of them, to one treatment with diflubenzuron. The number of studied generations are variable for every population, continuing it up to the moment in which already the presence was not observed anomalous forms.

Substrate preference by inmatures phases of blackflies (Diptera: Simulidae) in rivers of valencian autonomous region

D. López-Peña, A. Lis-Cantín and R. Jiménez-Peydró
Institut Cavanilles (Universitat de València), Laboratorio de Entomología y Control de Plagas, C/Catedrático José Beltrán Martínez, 2, 46980 Paterna (Valencia), Spain; david.lopez@uv.es

The Simuliidae family, is poorly studied in the Valencian autonomous region. This fact, together with the sanitary and veterinary importance of this group widely contrasted in the bibliography and the caused damage to the economic, tourism and livestock sector, as well as health and well-being animal and human due to bites suffered and possible parasitic diseases transmited. These have been the reason that has promoted the study in order to contribute to the increased knowledge of this group in the study latitudes. The aim of this study is to determine whether species of blackflies identified in a number of sampled rivers of the Valencian geography, show or not a preference in the choice of substrate at the time of the transition between the immature and adult or imago state. Is, see if the states larval and pupal of specific species show preference for one type of substrate where you can join or if on the contrary they aren't show any difference. The results show that species of blackflies identified show significant preference as to adhere to a substrate or another where complete part of their life development.

MOOC in medical entomology: free online course on insect vectors and related pathogens

V. Robert[1], M. Picard[1] and A.B. Failloux[2]
[1]IRD, MIVEGEC/IRD 224-CNRS 5290, Montpellier University, France, 911 Avenue Agropolis, 34394 Montpellier, France, [2]Institut Pasteur, Arbovirus and Insect Vectors, 25 rue du Dr Roux, 75724 Paris cedex 15, France; marie.picard@ird.fr

The understanding of interactions between the arthropods and the pathogens they may transmit is pivotal in combating many human and animal diseases. This angle has been adopted to support this new, open and interactive online course in medical and veterinary entomology, which will start on December 15, 2016. A MOOC is a Massive Open Online Course. It is a model for delivering learning content online to any person who wants to benefit of training, free of charge, with no limit on attendance. It is an important way for more people to access the high quality education that the Institut Pasteur and Institut de Recherche pour le Développement jointly deliver. This MOOC is closely related to the four-week Pasteur-IRD classes on insect vectors and pathogen transmission that is held every two years in April at the Institut Pasteur in Paris. The MOOC 'Medical entomology' aims at providing the understanding in medical and veterinary entomology at the university level. It will teach the role of vectors in the functioning of ecosystems and how to interrupt the vector transmission chain. It is organized from an entomological perspective, with each session devoted to a particular taxonomic group of insects or ticks, delivered in English by a team of experts from the Institut Pasteur, IRD and other well-known scientific institutions. The course is designed for anyone with an interest in learning about pathogens and vectors and is relevant to those involved with vector control teams, students taking a healthcare or science-related degree. The MOOC is organised in 7 modules spread over 6 weeks; each module is divided into several video classes. MCQs allow the learners to reassess their learning. The MOOC will be evaluated by weekly tests, video analysis, answers to small and simple exercises, as well as a final test. A certificate attesting successful completion of the course will be delivered. The Institut Pasteur, the CNAM and IRD co-produce this MOOC, run on the FUN platform. MOOC subscription started in September 2015 and the course will be launched on December 15, 2016.

Study of susceptibility/resistance status of insecticides in *Anopheles gambiae s.s.* (Giles, 1902) from two localities of Bouaké area (Côte d'Ivoire)

I.Z. Tia
Master International d' Entomologie (MIE), Bv 47, Abidjan, Côte d'Ivoire; tiazran@yahoo.fr

Malaria vector control strategies rely heavily upon the use of insecticide treated nets (ITNS) and indoor residual spraying (IRS). In Côte d'Ivoire, the National Malaria Control Programme (NMPC) strategies are based on effective case management and high coverage of populations with long lasting insecticidal nets. Unfortunately, the growing development of pyrethroid resistance constitutes a serious threat to malaria control programme and if measures are not taken in time, resistance may compromise control efforts in the foreseeable future. We conducted bioassays to inform NMPC of resistance status of the main malaria vector, *Anopheles gambiae s.s.* and the need for close surveillance of resistance. *Anopheles gambiae s.s.* from two localities of Bouaké area were collected in breeding sites and reared until emergence. Who susceptibility tests with impregnated filter papers were carried to detect resistance to three pyrethroids (deltamethrine, alphacypermethrin, permethrin) one carbamate and organophosphate with and without the inhibitor piperonyl butoxide (PBO) were carried to detect resistance status. The mortality rates of *Anopheles gambiae* with all insecticides without PBO were below 67% for both sites except pyrimiphos – methyl (organophosphate) with a mortality rate of 100% in the two localities. Pre-exposure to PBO highly increased these mortality rates to nearly 100%, indicating an involvement of oxidases and esterases in the resistance observed. Similar resistance level was observed for bendiocarb (carbamate) but PBO did not increased its activity in any of the mosquito populations. The results suggest that a package of resistance mechanism is present in this area of Côte d'Ivoire. News tools to manage these complex mechanisms are urgently needed.

First dengue serosurvey in Madeira Island: the real burden of the 2012 outbreak

A. Jesus[1], G. Seixas[1], T. Nazareth[1], A.C. Silva[2], R. Paul[3] and C.A. Sousa[1]
[1]GHTM-Instituto de Higiene e Medicina Tropical-UNL, Rua da Junqueira, 100, 1349-008 Lisboa, Portugal, [2]Departamento de Saúde, Planeamento e Administração Geral, IASAUDE, IP-RAM, Rua das Pretas, 1, 9004-515 Funchal, Portugal, [3]Functional Genetics and Infectious Diseases Unit, Institut Pasteur, 25-28 Rue du Docteur Roux, 75724 Paris, France; ana.spj20@gmail.com

Dengue fever is a vector-borne disease that has spread rapidly throughout the world. It is caused by a *Flavivirus* and transmitted by mosquitoes of the genus *Aedes*. There are an estimated 400 million infections per year, in more than 100 endemic countries, of which 80% are asymptomatic. Seven years after the introduction of *Aedes aegypti* (Linnaeus, 1762) in Madeira, a dengue outbreak was declared in October 2012 with a total of 2,168 probable cases of which 1,080 were confirmed. Thus, based on the threshold of 80% asymptomatic cases, the overall impact of the outbreak can be estimated in circa 5,400 human infections. Our study aims to perform the first seroprevalence survey in Madeira Island after the outbreak and assess the usefulness of saliva for dengue IgG detection with Panbio® kit. Serum and saliva samples were collected from 355 participants with ages over 10 years old. Anti-dengue IgG detection was carried-out with a commercial kit – IgG indirect ELISA, Panbio® and secondary infections determined by IgG Capture ELISA kit (Panbio®). Regarding serum samples, we observed 32 positive results. Positive samples were also tested with IgG Capture ELISA kit showing that 5 samples are positive for recent secondary infections. IgG indirect ELISA showed no promising results for dengue antibody detection in saliva samples. These results suggest an estimated 19,000 people were affected by 2012's dengue outbreak in Madeira Island, an estimate almost 8 times higher than the notified probable cases during the outbreak. The presence of positive samples that were probable secondary infections in individuals not previously exposed in endemic countries outside of Madeira suggests potential previous transmission of dengue in Madeira and highlights the importance of dengue surveillance and mosquito control strategies as a key factor to prevent future outbreaks. Funding: SFRH/BD/98873/2013; GHTM – UID/Multi/04413/2013; FP7-HEALTH-DENFREE (Ref. 282378).

Increased activity of sindbis virus in the enzootic vectors are potential neutral outbreak indicators

J.O. Lundström[1,2], J.C. Hesson[2], M. Schäfer[1], Ö. Östman[3] and M. Pfeffer[4]
[1]NEDAB, Swedish Biological Mosquito Control, Kölnavägen 25, 81021 Gysinge, Sweden, [2]Uppsala University, Medical Biochemistry & Microbiology/Zoonotic Science Centre, P.O. Box 582, 75123 Uppsala, Sweden, [3]Institute of Coastal Research/Swedish University of Agricultural Sciences, Aquatic Resources, Skolgatan 6, 74242 Öregrund, Sweden, [4]Animal Hygiene and Veterinary Public Health/University of LeipzigH, An den Tierkliniken 1, 04103 Leipzig, Germany; jan.lundstrom@mygg.se

Outbreaks of sindbis virus (SINV) infections, with polyarthritis and rash in humans, have been suggested to occur in a regular seven year cycle, based on the recorded annual number of cases in Sweden and Finland. The regularity of SINV-induced polyarthritis outbreaks is however not well defined, since reporting of cases is biased by multiple human factors. In search of a neutral marker of virus activity, we studied the SINV infection rates in vector mosquitoes in a Swedish endemic area during the pre-outbreak years 2000 and 2001, the hypothesized outbreak years 2002 and 2009, and the post-outbreak year 2003. We sampled mosquitoes in the River Dalälven floodplains, central Sweden, and the enzootic vectors *Culex torrentium/pipiens* and potential vector *Culiseta morsitans* were assayed for virus by cell culture. Partial nucleotide sequences of recovered virus strains were compared using neighbour joining and median joining network analysis. We recovered one SINV strain in 2001, 5 in 2002, 16 in 2009, and 0 in 2000 and 2003. The highest infection rates were observed in the enzootic vectors *Culex torrentium/pipiens* in 2002 and 2009, with 10.1/1000 and 21.0/1000, respectively. Our results show that both increased SINV prevalence in the enzootic vector *Culex torrentium/pipiens*, and the abundance of the enzootic vector, are potential neutral indicators of outbreak years. All isolates were shown by neighbour joining of structural protein genes to belong to the SINV genotype I. These findings suggest that increased SINV prevalence in enzootic vector mosquitoes, and abundance of these species, are reliable outbreak indicators, and that the virus remains in the endemic area for decades without evidence for frequent reintroductions of new virus strains.

Fine scale mapping of malaria infection clusters by using routinely collected health facility data in urban Dar es Salaam, Tanzania

Y.P. Mlacha[1,2], D.J. Terlouw[1], G.F. Killeen[1,2] and S. Dongus[1,2]
[1]Liverpool School of Tropical Medicine, Vector Biology Department and Department of Clinical Sciences, Liverpool, L3 5QA, United Kingdom, [2]Ifakara Health Institute, Environmental Health and Ecological Science, P.O. Box 78373, Kiko Avenue Mikocheni, 14112, Tanzania; ymlacha@ihi.or.tz

This study investigated whether passively collected routine health facility data can be used for mapping spatial heterogeneities in malaria transmission at the level of local government housing cluster administrative units in Dar es Salaam, Tanzania. From June 2012 to Jan 2013, residential locations of patients tested for malaria at a public health facility were traced based on their local leaders' names and geo-referencing these leaders' houses. Geographic information systems were used to visualize the spatial distribution of malaria infection rates, and for calculating Topographic Wetness Indices (TWIs). Spatial scan statistics were deployed to detect spatial clustering of high infection rates. Potential spatial associations of malaria infection rates with TWI were assessed by regression analysis. Among 2,407 patients tested for malaria, 46.6% (1,121) could be traced to their 411 different residential housing clusters. One small spatially aggregated cluster of neighborhoods with high diagnostic positivity was identified. High diagnostic positivity was associated with high maximum TWI. While the home residence housing cluster leader was unambiguously identified for 73.8% (240/325) of malaria-positive patients, only 42.3% (881/2,082) of those with a negative test result were successfully traced. Recording simple points of reference during routine health facility visits can be used for mapping malaria infection burden on very fine geographic scales, potentially offering a feasible approach to rational geographic targeting of malaria control interventions. However, in order to tap the full potential of this approach, it will be necessary to optimize patient tracing success and eliminate biases by blinding personnel to test results.

Overview on canine vector-borne diseases agents transmitted by ticks in Portugal

I. Pereira Da Fonseca[1], C. Marques[1], A. Duarte[1], A. Leal[2], J. Meireles[1], M.M. Santos-Silva[3] and A.S. Santos[3]
[1]CIISA, Faculty of Veterinary Medicine, University of Lisbon, Av. Universidade Técnica, 1300-477 Lisboa, Portugal, [2]Merial Portuguesa, Empreendimento Lagoas Park, Edificio 7 Piso 3, 2740-244 Porto Salvo, Portugal, [3]Centre for Vectors and Infectious Disease Research, National Institute of Health Dr Ricardo Jorge, 2965-575 Águas de Moura, Portugal; ifonseca@fmv.ulisboa.pt

Vector-borne diseases have a great animal/human health impact causing up to one million deaths/year. Ticks are important vectors of canine vector-borne diseases (CVBD) caused by *Babesia*, *Ehrlichia*, *Anaplasma*, *Rickettsia*, *Borrelia* and *Mycoplasma* genus, which may be zoonotic. Hence, companion animals may have an important role as reservoirs of such diseases to humans. In this work we aimed to do an update on CVBDs in Portugal by reporting data from several seroprevalence studies and from a FMV/CEVDI/Merial joint study about CVBD detection on ticks collected in this country. Ticks and CVBDs are widely disseminated in the Portuguese territory. In the last two decades, published seroprevalence values of CVBD agents have ranged between: 1.6-76.4% for *Babesia* spp., 23.8-85.6% for *Rickettsia conorii*, 4.1-50% for *Ehrlichia canis*, 4.5-54.5% for *A. phagocytophilum* and 0.2-38% concerning *B. burgdorferi s.s.*, *B. afzelii* and *B. garini*. In the FMV/CEVDI/Merial study, protozoan and bacterial pathogens were detected in ticks collected from owned dogs all over the country. This molecular study registered estimated proportions of tick infection of 10-65% for *Babesia*/*Theileria* spp.; 11-64% *Anaplasma*/*Ehrlichia* spp.; 0.54% *Coxiella burnetti* and 5-38% for *Rickettsia* spp. *Rhipicephalus sanguineus* was the most prevalent tick and all the above mentioned CVBD agents were detected within this tick species. *Rickettsia* spp. DNA sequencing revealed the presence of *R. massiliae* and *R. connori* in *R. sanguineus*. Furthermore, *R. sibirica* mongolotimonae was detected in *R. pusillus*. The data here presented highlights the dissemination of CVBD agents in dogs and ticks from Portugal. The early CVBD diagnosis and rigorous use of ectoparasite prevention by veterinarians/owners is crucial to prevent further dissemination of ticks and CVBDs – some of them with zoonotic implications – in a One Health aproach.

Distribution and seasonality of *Culicoides* species in Portugal (2005-2013)

D. Ramilo[1], S. Madeira[1], R. Ribeiro[1], T.P. Nunes[1], G. Alexandre-Pires[1], T. Nunes[2], F. Boinas[1] and I. Pereira Da Fonseca[1]
[1]CIISA, Faculty of Veterinary Medicine, University of Lisbon, Animal Health, Av. Universidade Técnica, 1300-477 Lisbon, Portugal, [2]Faculty of Sciences, University of Lisbon, Microscopy Unit, Campo Grande, 1749-016 Lisbon, Portugal; dwrramilo@hotmail.com

A National Entomological Surveillance Program (NESP) for bluetongue disease (BTD) was implemented in mainland Portugal and islands (Azores and Madeira), between 2005 and 2013, in order to obtain data on *Culicoides* as part of the Portuguese Official Veterinary Services program for the prevention and control of BTD. The objective was to evaluate *Culicoides* species distribution and seasonality in Portugal between 2005 and 2013. During the NESP, Portugal was divided into a grid of 57 geographic units in order to select the farms where *Culicoides* captures were carried out. Insects were collected using Centers for Disease Control (CDC) light traps and insect samples were analysed under stereoscope microscopy (SM). *Culicoides* species were identified using SM and, when necessary, specimens were prepared for composed optical microscopy and scanning electron microscopy. Several identification keys were used. Besides the 47 species previously known in Portugal, 21 species were identified for the first time, including vector species from Obsoletus group in all islands of Azores, as well as a novel species, *Culicoides paradoxalis*. *Culicoides* species showed different patterns of distribution and seasonality. *Culicoides* species have diverse ecological preferences justifying their different distribution in Portugal. It is also remarkable that several species can prevail all year, even during colder months, in a process called overwintering. Global warming and climate change may justify this condition and further studies must be conducted to understand if those species play a role in BT maintenance and dispersion. Vectors of BT are also present in both archipelagos, being the surveillance in these regions crucial. Funding: FCT projects UID/CVT/00276/2013, Pest-OE/AGR/UI0276/2014 and FCT grant SFRH/BD/77268/2011. Acknowledgments go to all the collaborators of NESP (DGAV/FMV).

Genetic diversity and population structure of *Aedes aegypti* from a trans-border region in Amazonia

P. Salgueiro[1], J. Pinto[1], A. Martins[2] and I. Dusfour[3]
[1]IHMT, GHTM, Rua da Junqueira, 100, 1349-008 Lisboa, Portugal, [2]Fiocruz-IOC, Av. Brasil, 4365, Manguinhos, Rio de Janeiro, RJ, 21040-360, Brazil, [3]Institut Pasteur de la Guyane, 23 av Louis Pasteur, 97300 Cayenne, France; psalgueiro@ihmt.unl.pt

In recent years South America has been afflicted by successive high impact outbreaks of dengue, chikungunya and zika. *Aedes aegypti* is the main mosquito vector of these diseases and the control of this mosquito is the sole solution to reduce transmission. In order to improve vector control it is essential to study mosquito population dynamics. Here we present a population genetics study in a trans-border region in Amazonia between Brazil and French Guiana to provide further knowledge on these questions. Two sites in French Guiana (SGO and CAY) were prospected for mosquito eggs in December 2013, May-Jun. 2014 and December 2014. Other two sites in the neighbor Amapá state in Brazil were also prospected in May-June 2014 (OIA and MAC). F0 adults from these collections were genotyped for 13 microsatellites to assess genetic diversity, population differentiation and effective population size. We also used a Bayesian clustering method to evaluate patterns of population structure. Levels of genetic diversity and effective population size were comparable among the populations studied. Pairwise genetic differentiation over temporal samples was six fold lower than differentiation over distinct spatial samples collected in the same season. None of the pairwise comparisons between temporal samples was significantly differentiated. The most differentiated sample was MAC in Brazil with all comparisons being significant. A significant correlation was found between genetic and geographic distances ($r=0.81$, $P=0.02$), suggesting isolation by distance. We detected the presence of 3 clusters ($k=3$) congruent with CAY, SGO+OIA and MAC. The estimated number of migrants among samples was consistent with genetic differentiation results. Results suggest: (1) high gene flow between the closest trans-border samples SGO in French Guiana and OIA in Brazil; and (2) barriers to gene flow between the detected clusters that may hinder the exchange of other genes of interest alleles (e.g. resistance alleles).

The *Anopheles gambiae* 2La chromosome inversion is genetically associated with susceptibility to *Plasmodium falciparum* throughout Africa

M.M. Riehle[1], T. Bukhari[2], A. Gneme[3], M. Guelbeogo[3], N. Sagnon[3], B. Coulibaly[4], A. Pain[2], E. Bischoff[2], F. Renaud[5], A.H. Beavogui[6], S.F. Traore[4] and K.D. Vernick[1,2]
[1]University of Minnesota, 1500 Gortner Ave, St Paul, MN 55108, USA, [2]Institut Pasteur, 28 Rue du Docteur Roux, 75015 Paris, France, [3]Centre National de Recherche et de Formation sur le Paludisme, BP 2208, 01 Ouagadougou, Burkina Faso, [4]Malaria Research and Training Centre, University of Mali, Point G, Bamako, 001, Mali, [5]MIVEGEC Maladies Infectieuses et Vecteurs: Ecologie, Génétique, Evolution et Contrôle, 911 Avenue Agropolis, 34394 Montpellier, France, [6]Centre de Formation et de Recherche en Sante Rurale de Maferinyah, BP 2649, Conakry 001, Guinea; kvernick@pasteur.fr

The *Anopheles gambiae* species complex contains multiple segregating chromosome inversions, which suppress recombination and lock together alternate haplotypes as supergenes. The large 2La inversion segregates in the two main African malaria vectors, *An. gambiae* and *An. coluzzii*, and is geographically widespread in Africa. We show that the inverted and standard allelic forms of the inversion are significantly associated with natural *Plasmodium falciparum* infection in wild-captured vectors from multiple locations in West and East Africa. Controlled experimental infection of wild mosquitoes with natural *P. falciparum* also reveals the same association with infection, indicating that the phenotype is at least partly due to differences in physiological susceptibility. The 2La inversion genotype is also associated with behavioral variation for indoor or outdoor preference. The 2La inversion is a major source of heterogeneity in the natural *P. falciparum* transmission system. Due to their elevated malaria susceptibility and exophilic tendency, some 2La inversion genotypes may generate persistent reservoirs of residual transmission in the malaria pre-elimination stage.

Identification and epidemiological implications of *Culicoides* biting midges in Trinidad, West Indies

T.U. Brown-Joseph[1], L.E. Harrup[2], V. Ramkissoon[1], R. Ramdeen[1], C.V.F. Carrington[1], S. Carpenter[2] and C.A.L. Oura[1]
[1]The University of the West Indies, Faculty of Medical Sciences, E.W.M.S.C., St. Augustine, Trinidad and Tobago, [2]The Pirbright Institute, Vector-borne Viral Disease Programme, Woking, GU24 0NF, United Kingdom; tamiko.brown-joseph@sta.uwi.edu

Culicoides biting midges (Diptera, Ceratopogonidae) are biological vectors of viruses affecting both livestock (bluetongue virus (BTV) and epizootic haemorrhagic disease virus (EHDV)) and humans (Oropouche virus). This study was carried out to identify which *Culicoides* species are present in Trinidad, the larger of the twin-island Republic of Trinidad and Tobago, which are the southern-most islands of the Caribbean chain adjacent to Venezuela. *Culicoides* abundance and distribution are strongly influenced by local-scale landscape factors; therefore, to develop a targeted sampling strategy remotely-sensed climate data was used to divide Trinidad into nine distinct ecozones. In this study, we present the *Culicoides* species found, and the epidemiological implications of these species, in three of the nine ecozones present in Trinidad, which are characteristic of three distinct habitat types (swamp, rural livestock farming and tropical forest). Insect collections were carried out bimonthly using miniature CDC light traps (BioQuip Products Inc.) placed overnight in six trap sites within each ecozone. Collected *Culicoides* were identified via both morphology, using established biological keys and molecular methods based on sequence comparison of the DNA barcode region of the mitochondrial cytochrome c oxidase I (COI) gene. To date ten *Culicoides* species have been identified via morphology. It was however not possible to identify to species-level for several specimens based on morphological or molecular methods, indicating that they may be novel species within the fauna of Trinidad. Within the three ecozones, the tropical forest revealed the highest species diversity with ten known and four unknown species identified to date. In contrast, there was limited species diversity identified on the rural livestock farm, where *C. insignis* Lutz dominated collections (96%). Interestingly, *C. insignis* is a known vector of BTV and EHDV, and these viruses have been confirmed to be highly prevalent in cattle at this location.

Ecology of tick-borne pathogens: investigation on rodents and ticks in protected areas in Central Italy

I. Pascucci[1], M. Di Domenico[1], G. Angelico[2], V. Curini[1], C. Cammà[1] and G. Sozio[1]
[1]IZS dell'Abruzzo e del Molise, Campo Boario, Teramo, Italy, [2]IZS dell'Umbria e delle Marche, Via Salvemini 1, Perugia, Italy; i.pascucci@izs.it

A study aimed to investigate the eco-epidemiological relationships among pathogens, ticks and rodents was performed in protected areas of Central Italy. *B. burgdorferi s.l.*, SFG Rickettsiae, *C. burnetii*, *A. phagocytophilum* were investigated. 24 sites were sampled by Sherman traps for rodents from July 2014 to April 2016. 128 blood samples belonging to *A. flavicollis* (93), *A. sylvaticus* (8), *M. glareolus* (24) and *E. quercinus* (3) were collected from the mandibular plexus of sedated rodents resulting positives for *B. burgdorferi s.l.* (4%). Feeding ticks were also collected and stored in 70% ethanol. One third of sites were also sampled by'dragging'.Ticks were identified by morphology and/or by *12S* gene sequencing. Moreover, a specific real time PCR for *I. ricinus* has been developed to distinguish between *I. ricinus* and *I. acuminatus*. Despite the morphological resemblance with *I. ricinus*, *I. acuminatus* is an endophilic rodents-associated tick, which life cycle and vectorial competence are poorly investigated. DNA was tested for pathogens by real time PCR. DNA sequencing was also performed on positives samples to identify pathogens at the species or geno-species level. From rodents, 315 feeding ticks (mainly *I. ricinus* 49.5% and *I. acuminatus* 37.5%) were collected and tested; *B. afzelii* and *R. raoultii/R. slovaca* and *R. monacensis* were found. *B. afzelii* prevalence in *I. acuminatus* was 32%,while only 1.1% in *I. ricinus*. 264 free-living ticks (mainly *I. ricinus)* were collected and pooled according to species and development stage.19 pools resulted positives for *B. burgdorferi s.l.* (4 *B. valaisiana;* 6 *B. garinii;* 9 *B. afzelii*) and 18 for *Rickettsia* spp. (9 *R. monacensis*, 1 *R. slovaca*, 8 *Rickettsia* spp.). Our results provide new data on wild rodents' role in maintaining the agents of Lyme disease and Rickettsiosis in the studied area, suggesting that *B. afzelii*, the human pathogenic rodent associated geno-species, is likely to be maintained at rodents' level by the specialist tick *I. acuminatus*. Detection of other geno-species associated to ground feeding birds in free living ticks confirms the complexity of the eco-epidemiology of Lyme disease in the Mediterranean area.

Dominant role of *Anopheles funestus* Giles, in a residual transmission setting in south-eastern Tanzania

E.W. Kaindoa[1,2], N.S. Matowo[1,2], F.C. Meza[2], H.S. Ngowo[2] and F.O. Okumu[1,2]
[1]University of the Witwatersrand, School of Public Health, Johannesburg, South Africa, South Africa, [2]Ifakara Health Institute, Environmental Health and Ecological Sciences, Off Mlabani Passage, 53, Ifakara, Tanzania; ekaindoa@ihi.or.tz

Malaria is transmitted by more than one anopheline mosquito species, but the roles of these species vary in different epidemiological settings. We assessed the role of the two dominant *Anopheles* mosquito species in selected villages in Ulanga district in south-eastern Tanzania. Mosquito sampling was done in randomly selected households in three villages using CDC light traps, back pack aspirators and HLC between January 2015 and January 2016. Multiplex polymerase chain reaction was used to identify members of the *An. funestus* group mosquitoes and *An. gambiae s.l.*, to species level. Enzyme-linked immunosorbent assay was done to detect *Plasmodium* sporozoites in the mosquito salivary glands, and to identify sources of mosquito blood meals. Insecticides susceptibility tests were also conducted as per WHO guidelines. A total of 20,260 *An. arabiensis* and 4,802 *An. funestus* were collected. Among the *An. funestus* group mosquitoes, *An. funestus s.s.* predominated (76.6%), *An. rivulorum* (2.9%) and *An. leesoni* (7.1%) and unamplified samples (13.4%). About 86% of all infected mosquitoes were *An. funestus s.s.* while 14% were *An. arabiensis.* Overall, *An. funestus* group contributed to 93.4% and *An. arabiensis* contributed to 5.6% respectively of the annual EIR. In the *An. funestus* group, *An. funestus s.s* contributed to 96% of the transmission while *An. rivulorum* contributed 4%. Mosquito blood meal sources included: humans, 79.2%, bovine, 17.1%, dog, 2.4% and chicken, 1.2%. The findings also showed that *An. funestus* were resistant to Permethrin, Deltamethrin, and Lambda cyhalothrin. The evidence from this study demonstrates that *An. funestus* have a significant role as the main driver of malaria transmission in these study villages. The findings also shows high level of insecticides resistance in *An. funestus* species to various insecticides commonly used for vector control. The study underlines the importance of species-specific targeted interventions so as to reduce malaria transmission levels.

Borrelia burgdorferi s.l. infection in *Ixodes ricinus* ticks (Ixodidae) at three Belgian nature reserves: increasing trend?

I. Deblauwe[1], T. Van Loo[1], L. Jansen[1], J. Demeulemeester[1], K. De Witte[1], I. De Goeyse[1], C. Sohier[1], M. Krit[1] and M. Madder[2]
[1]Institute of Tropical Medicine, Biomedical sciences, Nationalestraat 155, 2000 Antwerp, Belgium, [2]University of Pretoria, Veterinary Tropical Diseases, Private Bag X04, Onderstepoort 0110 Pretoria, South Africa; tvanloo@itg.be

The incidence of the most prevalent tick-borne infection of humans in Europe, lyme borreliosis, increased in several European countries these last decades. It is caused by *Borrelia burgdorferi s.l.* and is transmitted by the sheep tick *Ixodes ricinus*. A longitudinal study at three Belgian nature reserves was performed to investigate the *B. burgdorferi s.l.* infection in *I. ricinus*. Ticks were collected by flagging in April and May from 2008 until 2016 at two forest nature reserves in the province of Limburg (the 'Wik' and the 'Ziepbeekvallei') and in all months from 2013 until 2016 at the dune nature reserve the 'Westhoek' in the province of West-Flanders. Collected ticks were pooled and screened for *B. burgdorferi s.l.* using a semi-nested PCR. Tick questing activity (average number per minute) increased the first years at the 'Wik' and the 'Ziepbeekvallei', but respectively, stabilised and decreased in 2014 and 2016. The increased hunt on wild boars at both reserves and the forestry work at the 'Ziepbeekvallei' probably can explain the stopped increase in tick activity. At the 'Westhoek' the questing tick activity (average number per sampling day) increased in 2015 and 2016. The individual infection rate of *I. ricinus* with *B. burgdorferi s.l.* at the 'Wik' was stable until 2014 (average = 10%, n = 5 years, range 6%-13%), but increased threefold in 2016 (37%). At the 'Ziepbeekvallei' the sample size of screened nymphs was in most years too small to draw conclusions. As at the 'Wik', the individual infection rate of *I. ricinus* at the 'Westhoek' increased between 2014 and 2016 from 13% to 24%. The warmer winters of 2014 and 2016 together with the increasing frequency of mast years might be linked to the increase of tick questing activity and infection rate. Host reservoirs (mainly rodents) probably increased in numbers and were active longer and as such more tick larvae could feed, survive and get infected. A more profound study is needed to confirm this hypothesis.

Eco-entomological features associated with the presence of triatomines in Northwest Córdoba, Argentina

C. Rodriguez, A. Lopéz, J. Nattero, C. Soria, M. Cardozo, P. Rodriguez and L. Crocco
Instituto de Investigaciones Biológicas y Tecnológicas (IIBYT-CONICET/UNC), FCEFyN, UNC, Av. Velez Sarsfield 1611, Córdoba, 5016, Argentina; paula.ro.san55@gmail.com

Chagas disease is a Latin-American zoonosis with a high historical endemism in the Gran Chaco region; particularly in Argentine, this area occupies northwest and center of the country. *Triatoma infestans* is the main vector of this disease in Argentina, and their presence in human habitats depends on environmental, geographical, cultural and social aspects. The work was conducted in 139 household units (HU: dwelling and associated peridomicile) from a define rural area of center Argentina (2012-2013). At each HU, Triatomines were captured, risk factors and the kind of PA (chicken coops, goat or pig corrals) were registered. Interviews were done to inhabitants to obtain demographic data, knowledge, attitudes and practices regarding this disease. Results were classified by the number of correct answers in 3 levels: I: 1 to 3, II: 4 to 6 and III: 7 or more than 7. Entomological indicators were calculated: domiciliary and peridomiciliary infestation index (DI and PI), household unit infestation rate (HUI). The data was analyzed using descriptive statistics, crosstabs, odds ratios (OR) and confidence intervals (95% CI). From the 1,135 Triatomines collected, 98% were *T. infestans*. The HUI was 59.7%, DI 4.3%, PI 58.3% and 26% of dwellings were colonisable. The 80% of dwellings with risk walls and 60% with risk roofs were positive for Triatomines, of which 95% had chicken coops and were 12.03 times riskier than those without (p=0.005). PAs were in 85.6% of the HU visited; being chicken coops the most frequent (78.4%). The 57.5% of chicken coops was less than 12 m away from dwellings, of which 50% were infested with triatomines. From the 95 residents surveyed, 66.3% were at Level II and 24% had Chagas disease. Most of them heard about this disease and identified *T. infestans*. The 63%, 86.5% and 93.3% of them were unaware of treatment, congenital and transfusional transmission, respectively. Our results showed that inhabitants do not perceive peridomiciles as an area that favors the presence of vectors. Transfer actions to identify factors that favor the presence of triatomines are needed to improve peridomiciliary area.

Borrelia miyamotoi and co-infection with Borrelia afzelii in Ixodes ricinus ticks and rodents from Slovakia

Z. Hamšíková[1], C. Coipan[2], L. Mahríková[1], H. Sprong[2] and M. Kazimírová[1]
[1]Institute of Zoology, Slovak Academy of Sciences, Dubravska cesta 9, 84506 Bratislava, Slovak Republic, [2]National Institute for Public Health and Environment, Laboratory for Zoonoses and Environmental Microbiology, 9 Antonie van Leeuwenhoeklaan, P.O. Box 1, Bilthoven, the Netherlands; maria.kazimirova@savba.sk

Borrelia miyamotoi is a tick-borne pathogen that causes relapsing fever in humans. The occurrence of this spirochaete has recently been reported in *Ixodes ricinus* ticks and rodents from a number of European countries, but there are still gaps in the knowledge of its eco-epidemiology and impact on public health. In the current study, questing *I. ricinus* nymphs and adults as well as skin biopsies from wild-living rodents captured in Slovakia were screened for the presence of *B. miyamotoi* and *B. burgdorferi s.l.* DNA by a triplex real-time polymerase chain reaction, targeting ospA and flagellin genes. The overall prevalence of *B. miyamotoi* and *B. burgdorferi s.l.* in questing ticks was 1.7% and 21.3%, respectively. *Borrelia miyamotoi* was detected in the most abundant rodent species: *Apodemus flavicollis* (9.3%) and *Myodes glareolus* (4.4%). In contrast, *B. burgdorferi s.l.* was identified in 11.9% of rodents, with the highest prevalence (68.4%) in *Microtus arvalis*. Its prevalence in *Apodemus* spp. and *M. glareolus* was lower: 8.4% and 12.4%, respectively. *Borrelia afzelii* was the prevailing genospecies infecting questing *I. ricinus* (61.1%) and rodents (92.9%). Co-infections with *B. burgdorferi s.l.* and *B. miyamotoi* were found in 24.1% and 9.3% of the questing ticks and rodents, respectively. Associations of *B. miyamotoi* with *B. afzelii* were revealed, suggesting that the spirochaetes share amplifying hosts. A single *B. miyamotoi* genotype was identified in *I. ricinus* ticks and rodents captured during three consecutive years, indicating that this spirochaete is endemic in the area and rodents are involved in its enzootic cycle. The sequences of the *B. miyamotoi glpQ* gene fragment from our study showed a high degree of identity with sequences of the gene amplified from ticks and human patients in Europe. The results indicate that humans in Slovakia are at potential risk of contracting tick-borne relapsing fever, and in some cases together with Lyme borreliosis.

Identification of relapsing fever *Borrelia* in hard ticks from Portugal

M. Nunes, R. Parreira, T. Carreira and M.L. Vieira
Global Health and Tropical Medicine (GHTM), Instituto de Higiene e Medicina Tropical (IHMT), Universidade Nova de Lisboa (UNL), Rua da Junqueira, 100, 1349-008 Lisboa, Portugal; monicanunes@ihmt.unl.pt

Tick-borne pathogens are believed to be responsible for over 100,000 cases of human disease. Ticks are second to mosquitoes as worldwide vectors of human pathogenic agents, but the most important vectors of those of animals. These pathogens, include *Borrelia* species from *B. burgdorferi sensu lato (B.b.*s.l), responsible for Lyme disease (LD), relapsing fever (RF) and reptile-associated (REP) complexes. LD and REP are usually transmitted by ixodid ticks, while most RF are transmitted by argasid ticks. Lately some RF species including *B. theileri*, *B. miyamotoi*, and *B. lonestari* have adapted to ixodid ticks, such as *Rhipicephalus*, *Ixodes* or *Amblyomma*, that can act as their vectors. Questing ticks collected from eight Portuguese districts, were identified to species level, and their DNA extracted by alkaline hydrolysis. Four genes of *Borrelia* sp. were targeted for PCR: 16S rDNA, the intergenic 5S-23S spacer, *flaB*, and *glpQ*, followed by phylogenetic analysis of their sequences. The collected ticks were classified as *Dermacentor*, *Haemaphysalis*, *Hyalomma*, *Ixodes*, and *Rhipicephalus*. *B.b.s.l.* and RF *Borrelia* DNA were identified in the analyzed ticks, and the phylogenetic analysis indicated an association with RF species, such as *B. lonestari*, *B. theileri*, and unknown *Borrelia* detected in Japan and Brazil; and one sequence clustered with *B. miyamotoi* found in other European countries. Besides the presence of LD agents, two putative and uncharacterized RF-like *Borrelia* associated to hard ticks, and *B. miyamotoi* were also identified. Although the potential risk of the new RF-like *Borrelia* to humans is still unknown, further studies are necessary, including culture isolation, to understand and classify this species, and assess its distribution across Portugal. Human infections with *B. miyamotoi* have been reported in Russia, USA and the Netherlands, but not yet in Portugal. However, since ticks could carry both *B. miyamotoi* and LD spirochetes, increasing human co-infection risk, additional studies must be undertaken. Funding: study support by Fundação para a Ciência e a Tecnologia, for funds to GHTM – UID/Multi/04413/2013, and through a PhD grant (SFRH/BD/78325/2011).

Vector species of *Culicoides* implicated in bluetongue epidemics in Italy

M. Goffredo, G. Mancini, A. Santilli, V. Federici, M. Quaglia, F. Di Nicola, M. Catalani, L. Teodori and G. Savini
Istituto Zooprofilattico Sperimentale Abruzzo e Molise, Animal Health/Entomology, Campo Boario, 64100 Teramo, Italy; m.goffredo@izs.it

Since 2000 several incursions of bluetongue virus (BTV) occurred in Italy. In 2012 serotypes 1 and 4 of BTV entered and co-circulated in Sardinia. Then BTV-1 spread throughout Sardinia and invaded Sicily and the Italian Tyrrenian coast. In 2014, it spread extensively in mainland Italy, causing severe outbreaks. In late 2014 BTV-4 was detected in Southern Italy and, in 2015, both serotypes circulated in Central and Southern Italy. The National Entomological Surveillance Plan for BT includes *Culicoides* collections on BTV affected farms, as demonstrated by seroconversions in sentinel animals or presence of clinical outbreaks, with the aim of identifying the vector species involved in virus transmission. During the recent bluetongue epidemics occurred in Italy described above, an extensive entomological survey was conducted. *Culicoides* were collected on nearly 350 affected farms of 12 Italian regions, and about 3,000 pools (composed by more than 80,000 midges) were tested for the presence of BTV. They were composed by Obsoletus complex (including *C. obsoletus, C. scoticus* and *C. montanus)*, *C. imicola*, *C. newsteadi*, Pulicaris complex, *C. pulicaris*, *C. punctatus*, *C. dewulfi* and Nubeculosus complex. All these taxa resulted positive to BTV, at least once, with a total minimum infection rate (MIR) of over 1%. This study confirms that *C. imicola* and Obsoletus complex are the most important vectors of bluetongue in Italy. However, it clearly shows that other species may have played a role in transmitting the virus during the recent BTV epidemics, including *C. pulicaris*, *C. dewulfi* and some species never implicated before. In particular the RNA of BTV was repeatedly found in parous females of *C. newsteadi* and *C. punctatus*, collected on farms where BTV was circulating. Serotype 1 was detected in all tested taxa, whereas the BTV-4 was detected in the Obsoletus complex, *C. imicola*, and *C. newsteadi*. In addition, BTV-1 and BTV-4 were simultaneously found in pools of *C. imicola* and *C. newsteadi*. Interestingly, viral RNA was found also in some pools composed by not pigmented females. The possible epidemiological meaning of these findings is discussed.

Reversal and fitness costs of deltamethrin resistance in *Aedes aegypti* populations from French Guiana

N. Habchi-Hanriot[1,2], E. Ferrero[2], P. Gaborit[2], J. Issaly[2], A. Guidez[2], L. Wang[2], Y. Epelboin[2], R. Carinci[2], N. Pocquet[3], M. Pol[3], R. Girod[2] and I. Dusfour[2]
[1]International Master of Entomology (MIE), UM, France & UAO, Cote d Ivoire, [2]Institut Pasteur de la Guyane, Unité d'Entomologie Médicale, 23 avenue Pasteur, 97 306 Cayenne, French Guiana, [3]Institut Pasteur de Nouvelle-Calédonie, Unité d'Entomologie Médicale, 9-11 avenue Paul Doumer, 98 845 Nouméa, New Caledonia; nhabchihanriot@terra.com

Aedes aegypti is the vector of yellow fever, dengue, chikungunya and zika viruses in French Guiana. Considering the lack of vaccines against these viruses, a control of mosquitoes has been organized. Mechanical removal or treatment of breeding sites are performed against larvae and deltamethrin, the last adulticide allowed, is sprayed against adults. The massive use of deltamethrin since 2011 has leaded to the development of resistance, and then to a loss of the insecticide efficacy to control mosquitoes. In our project, reversal of resistance has been investigated in five laboratory strains, by stopping deltamethrin pressure either alone or in combination with introgression by the susceptible strain New Orleans. Fitness costs have also been characterized. The absence of insecticide pressure alone did not bring a substantial reversal. However, coupling the introgression process brought back susceptibility similar to New Orleans as early as the third generation. Considering the fitness, neither the mortality of immature stages nor the sex ratio has shown significant differences according to the level of deltamethrin resistance. However, resistant and introgressed strains presented high growth rates of immature stages, compared to the susceptible reference strain. Moreover, the introgressed mosquitoes were engorged at the same level of the susceptible strain whereas resistant mosquitoes took less blood meals and laid less eggs than the New Orleans. In addition, introgressed mosquitoes had a significantly higher fecundity than the resistant strains. The island strain IR, less resistant than the land-based ones, performed between the reference strain and the resistant ones, with an immature growth rate similar to New Orleans but lower gorging rate and fecundity. Therefore, fitness costs on local *Aedes aegypti* mostly include lower engorgement rate and fecundity.

Unchanged Schmallenberg virus seroprevalence in the Belgian sheep population after the season of 2014 and 2015 despite evidence of virus circulation

C. Sohier[1], R. Michiels[1], E. Kapps[1], E. Van Mael[2], C. Quinet[3], B. Cay[1] and N. De Regge[1]
[1]CODA-CERVA (Veterinary and Agrochemical Research Centre), Enzootic and (re)emerging diseases, Groeselenberg 99, 1180 Bruxelles, Belgium, [2]Dierengezondheidszorg Vlaanderen (DGZ), Deinse Horsweg 1, 9031 Drongen, Belgium, [3]Association Regionale de Santé et Identification Animalens (ARSIA), Allée des Artisans, 5590 Ciney, Belgium; charlotte.sohier@coda-cerva.be

The unexpected outbreak of Schmallenberg virus (SBV) that has occurred in Europe since 2011 has led to increased interest in *Culicoides* biting midges (Diptera: Ceratopogonidae). *Culicoides* which are already known to spread bluetongue virus and Akabane virus have been proposed to be the putative vectors of SBV. SBV induces only mild symptoms in adult ruminants but was shown to be responsible for abortions, stillbirths and congenital malformations in cattle, sheep and goats. Little data is available on the current antibody protection rate against SBV. Since the beginning of 2013, no more cross sectional seroprevalence studies have been performed to assess the protection rate of Belgian ruminants against SBV. Interestingly, three SBV suspected aborted calves that were submitted to the Belgian reference laboratory tested positive for SBV by qRT-PCR in April 2016, providing the first evidence of SBV circulation in Belgium since three years. A cross-sectional seroprevalence study was therefore carried out in the Belgian sheep population and showed that the total seroprevalence against SBV was 26% at the end of the vector season of 2015, being significantly lower than the seroprevalence of 84% detected after the outbreak in 2011. Nevertheless, 63% of the Belgian sheep flocks still had a certain level of protection against SBV. Despite the fact that PCR detection of SBV in aborted calves in April 2016 evidenced that SBV had circulated in 2015, no change in seroprevalence between 2014 and 2015 was found in the Belgian sheep population.

Towards high-throughput identification of arthropod vectors by mass spectrometry

C. Silaghi[1], V. Pflüger[2], P. Müller[3] and A. Mathis[1]
[1]National Center for Vector Entomoloy, Institute of Parasitology, 8057 Zürich, Switzerland, [2]Mabritec SA, Lörracherstrasse 50, 4125 Riehen, Switzerland, [3]Swiss TPH, Socinstrasse 57, 4002 Basel, Switzerland; cornelia.silaghi@uzh.ch

Accurate and high-throughput identification of vector arthropods is of paramount importance in surveillance and control programmes. Protein profiling by matrix-assisted laser desorption/ionization time of flight mass spectrometry (MALDI-TOF MS) technically fulfils these requirements, and reference databases have been established for several vector taxa by various research groups. However, approaches vary in terms of sample processing, equipment, data acquisition and analysis. Furthermore, these databases were typically generated on a project-by-project base, stored on local drives ('in-house databases') and thus are not accessible to the public. We have established the largest database with regard to vector arthropods, including Ceratopogonidae, Culicidae, Ixodidae, Phlebotominae and Simuliidae, and continually expand it in collaborative work, with the goal of creating a comprehensive, centralized database comprising reference spectra of all major arthropod vectors. The database is maintained by a private company ensuring both customer-oriented service and greater sustainability. Further, we for the first time showed that spectra obtained on mass spectrometers from different companies can be analysed using this database. In order to render the MALDI-TOF MS technology universally useful and available for the identification of arthropod vectors, we are developing harmonised protocols, including written and illustrated material as well as short film sequences, to provide instructions on arthropod handling and processing as well as on data acquisition on different instruments. Thus, mass spectra of arthropod vectors can be created on any MALDI-TOF MS machine and species identification obtained by electronically submitting the data to Mabritec SA for comparison with the established centralised database. The ultimate aim is an on-line database with access to the arthropod module for trained users. Ideally, a public core-funding would assure free access for non-commercial institutions and maintenance of the database (with regard e.g. to continued expansion, taxonomic adaptations).

The epidemiology of residual malaria transmission and infection burden in an African city with high coverage of multiple vector control measures

D. Msellemu
Ifakara Health Institute, EHES, P.O. Box 78373, Dar es Salaam, 14201, Tanzania; dmsellemu@ihi.or.tz

In the Tanzanian city of Dar es Salaam, high coverage of long-lasting insecticidal nets (LLIN), larvicide application (LA) and mosquito-proofed housing, was complemented with improved access to artemisinin-based combination therapy and rapid diagnostic tests by the end of 2012. Three rounds of cluster-sampled cross-sectional surveys of malaria parasite infection status, from 2010 to 2012, were complemented by two series of high-resolution, longitudinal surveys of vector density. Larvicide application by granule of *Bacillus thuringiensis* var. *israelensis* (Bti) had no effect upon either vector density (P=0.82) or infection prevalence (P=0.32) when managed by a private-sector. Infection prevalence rebounded back to 13.8% in 2010, from 11.8% compared to Bti LA evaluation in 2008. When the Ministry of Health and Social Welfare (MoHSW) managed LA, vector densities reduced, first by the same Bti granule in 2011 [odds ratio (OR) (95% confidence interval (CI) = 0.31 (0.14, 0.71), P=0.005] and then a pre-diluted aqueous suspension from mid-2011 onwards [OR (95% CI) = 0.15 (0.07, 0.30), P<0.001]. While LA by MoHSW with the granule formulation was associated with reduced infection prevalence [OR (95% CI) = 0.26 (0.12, 0.56), P<0.001], liquid suspension use, following a mass distribution was not (P=0.836). Sleeping inside houses with complete window screens reduced infection risk [OR (95% CI) = 0.71 (0.62, 0.82), P<0.001] Also, infection risk was only associated with local vector density [OR (95% CI) = 6.99 (1.12, 43.7) at one vector mosquito per trap per night, P=0.04] in few (14%) of households lacking screening. Despite attenuation of malaria transmission and immunity, 88% of infected residents experienced no recent fever, only 0.4% of these afebrile cases had been treated for malaria, and prevalence remained high (9.9%) at the end of the study. While existing vector control interventions have attenuated malaria transmission in Dar es Salaam, further scale-up and additional measures to protect against mosquito bites outdoors are desirable. Accelerated elimination of chronic human infections persisting at high prevalence will require active, population-wide campaigns with curative drugs.

Assessment of euthanasia as control measure of canine leishmaniasis using a longitudinal stochastic model

A. Muniesa[1], I. Ruiz-Arrondo[1,2] and I. De Blas[1]
[1]Faculty of Veterinary Sciences, Instituto Agroalimentario de Aragón (Universidad de Zaragoza-CITA), Miguel Servet 177, 50013, Zaragoza, Spain, [2]Center of Rickettsiosis and arthropod-borne diseases, Center for Biomedical Research of La Rioja, Piqueras 98, 26006, Logroño (La Rioja), Spain; irarrondo@riojasalud.es

Mathematical modelling is a powerful tool that allows simulating different scenarios, and it can be useful to develop different strategies for eradication and disease control. This kind of tools has advantages and disadvantages, it is really difficult to build a model that accurately represents the real world, because the algorithm can be so complex or so long to compute it, however, if we simplify the model to build it easier, it may not represent the real word. In this work we applied a stochastic model for canine leishmaniasis based on Monte Carlo method that calculate chances for each individual state change, using Bayesian probabilities and conditional logistic regression every two weeks. Dog variables (age, sex, purpose, health status), vector population and temperature have been considered. In some countries the use of euthanasia of infected dogs as control measure is a controversial issue. Our aim is to assess its effectiveness considering the conditions of our model (Mediterranean area, 5-years simulation period and 35% of initial seroprevalence). One of the coefficients of the model defines the quarterly probability to euthanise an infected dog (P) and it is conditioned by their purpose. The reference model assumed a P from 0.3% to 1.5%; so, the probabilities accumulated during a year (Pyear) can vary between 8.0% and 30.1%. The first scenario was based on P=5.7% for all purposes (Pyear=75.4%), and the second one set a probability almost null (P=0.015%, Pyear=0.299%). Overall seroprevalence is maintained during the study period in the reference model, but a small seasonality is observed. A relevant increment of euthanasia drives to a marked decrease of overall seroprevalence below 15% in 1 year, 10% in 2 years and later it remains around 3%. On the other hand, restriction of euthanasia goes to a continuous increment of seroprevalence over time. In conclusion, euthanasia could be an effective tool to control canine leishmaniasis, but it should be complemented with other measures to eradicate it.

High infectivity of *Anopheles melas* to *Plasmodium* in Southern Benin: implications for malaria transmission

C. Adigbonon, B.S. Assogba, L. Djossou and L.S. Djogbénou
Institut Régional de Santé Publique, Université d'Abomey-Calavi, 01BP918, Cotonou, Benin; adigbononclaudiane@gmail.com

Malaria is a worldwide disease affecting many people particularly in the tropical and sub-tropical areas. It is caused by *Plasmodium* parasites and essentially transmitted by female mosquitoes belonging to *Anopheles* genus. Our understanding of the infectivity of these vectors to *Plasmodium* is necessary design sustainable strategies for their control. This aspect remains unknown in the coastal and lagoon area of Benin where *An. melas* and *An. coluzzii* are sympatric. This study aims to investigate the infectivity of these two vectors to *Plasmodium* in order to understand their role in the transmission of malaria in coastal lagoon areas of Benin. Insecticides spray catches technique was used to collect females in 80 houses randomly selected in our study site. Three hundred and twenty females were identified using PCR–species technique and *Plasmodium* infection was determined by the TaqMan method during dry season. This assay detects all four malaria-causing *Plasmodium* species and discriminates *P. falciparum* from *P. ovale*, *P. vivax* and *P. malariae* (*OVM*). During the dry season, the sporozoïte rates were 0.2% and 0.3% for *An. melas* and *An. coluzzii*, respectively. However, we observed that positivity to the OVM (one of *Plasmodium ovale*, *Plasmodium vivax* and *Plasmodium malariae* species) was significantly higher in *An. melas* (95%) than in *An. coluzzii* (33.33%) (Chi-sq=15 857, df=1, $P<0.001$). These results indicated that *An. melas* is more infected by one of the species *P. ovale*, *P. vivax* and *P. malariae* than by *P. falciparum*, contrarily to *An. coluzzii*. These findings reinforce the debate on the role of *An. melas* in malaria transmission in coastal lagoon areas of Benin.

Seroprevalence of spotted fever group rickettsiae among a group of pet dogs from Luanda, Angola

P.F. Barradas[1], R. De Sousa[2], I. Amorim[1,3,4], H. Vilhena[5,6,7], F. Gärtner[1,3,4], L. Cardoso[8], A. Oliveira[9], S. Granada[9] and P. Silva[1]
[1]Institute of Biomedical Sciences Abel Salazar, University of Porto, Rua Jorge Viterbo Ferreira, Porto, Portugal, [2]National Institute of Health Dr Ricardo Jorge, Av. da Liberdade, 5, Águas de Moura, Portugal, [3]IPATIMUP, Rua Júlio Amaral de Carvalho, Porto, Portugal, [4]i3S, R. Alfredo Allen, Porto, Portugal, [5]EUVG, Coimbra, Lordemão, Coimbra, Portugal, [6]CECAV, UTAD, Quinta de Prados, Vila Real, Portugal, [7]Baixo Vouga Veterinary Hospital, Estrada Nacional 1, Segadães, Águeda, Portugal, [8]School of Agrarian and Veterinary Sciences, UTAD, Quinta de Prados, Vila Real, Portugal, [9]Casa dos Animais Veterinary Clinic, Morro Bento, Luanda, Angola; rita.sousa@insa.min-saude.pt

Dogs are important sentinels to identify tick-borne pathogens circulating in a specific geographic region. They are hosts of ticks and can be used to assess which tick-borne pathogens can infect them. Despite being important causes of systemic febrile illnesses in travellers returning from some countries from Africa, little is known about the presence of *Rickettsia* species circulating in dogs in Angola. Since we aimed to assess the importance of the dogs as hosts and carriers of ticks and tick-borne pathogens we have conduct a serosurvey for rickettsiae in a group of pet dogs from Luanda, Angola that were presented to consultation during January and February of 2013. From a total of 103 serum samples tested by in-house Immunofluorescence assay, six (5.8%) dogs showed the presence of reactive IgG antibodies against *R. africae.* All the dogs had access to the outside and 83% of them were infested with ticks at the time of blood collection or had history of tick infestation. No specific predisposing breed or crossbreed was found dogs and 83% of them were males. The age ranged between one and eight years and 67% of these animals have a canine tick-borne disease. The seroprevalence found in this study is much lower when compared with the seroprevalence found in other studies from European dogs. This fact could be related with the difference between the tick species that parasitize the dogs and harbour *Rickettsia* species.

Visualization of house-entry behaviour of malaria mosquitoes

J. Spitzen[1], T. Koelewijn[1], W.R. Mukabana[2] and W. Takken[1]
[1]Wageningen University, Laboratory of Entomology, P.O. Box 16, 6700 AA Wageningen, the Netherlands, [2]University of Nairobi, School of Biological Sciences, P.O. Box 30197 GPO Nairobi, Kenya; jeroen.spitzen@wur.nl

Malaria mosquitoes often blood feed indoors on human hosts. The mosquitoes predominantly enter houses via open eaves. Host-seeking is odour-driven, and finding a host depends on the quality of the odour plume and whether the route towards the host is free of obstructions. Little is known about in-flight behaviour of mosquitoes during house entry. This study visualizes mosquito house entry in three dimensions (3D) and offers new insights for optimizing vector control interventions. The approach and house entry of *Anopheles gambiae sensu stricto* was studied in a semi-field set-up using video-recorded flight tracks and 3D analysis. Behavioural parameters of host-seeking female mosquitoes were visualized with respect to their position relative to the eave as well as whether a mosquito would enter or not. Host odour was standardized using an attractive synthetic blend in addition to CO_2. The study was conducted in Kenya at the Thomas Odhiambo Campus of the International Centre of Insect Physiology and Ecology, Mbita. The majority of host-seeking *An. gambiae* approached a house with a flight altitude at eave level, arriving within a horizontal arc of 180°. Fifty-five per cent of mosquitoes approaching a house did not enter or made multiple attempts before passing through the eave. During approach, mosquitoes greatly reduced their speed and the flight paths became more convoluted. As a result, mosquitoes that passed through the eave spent more than 80% of the observed time within 30 cm of the eave. Mosquitoes that exited the eave followed the edge of the roof (12.5%) or quickly re-entered after exiting (9.6%). Host-seeking mosquitoes, when entering a house, approach the eave in a wide angle to the house at eave level. Less than 25% of approaching mosquitoes entered the house without interruption, whereas 12.5% of mosquitoes that had entered left the house again within the time of observation. Advances in tracking techniques open a new array of questions that can now be answered to improve (household) interventions that combat mosquito-borne diseases.

Community analysis on the diversity of mosquito and midge vector species across habitats in Europe

T.W.R. Möhlmann[1,2], L. Bracchetti[3], C. Damiani[3], G. Favia[3], W. Takken[2], M. Tälle[1], U. Wennergren[1] and C.J.M. Koenraadt[2]
[1]Linköping University, Valla 27, Linköping, Sweden, [2]Wageningen UR, Laboratory of Entomology, Droevendaalsesteeg 1, Wageningen, the Netherlands, [3]University of Camerino, Via Emidio Pacifici Mazzoni 2, Camerino, Italy; tim.mohlmann@wur.nl

Studies on European vector diversity often focus on a specific habitat, region or country, using a wide range of trap types. Results of such studies are therefore not harmonized and difficult to compare. A European-wide study comparing multiple countries and habitats could increase our knowledge about vector species diversity. We therefore collected monthly data on mosquito (Culicidae) and biting midge (*Culicoides* spp. communities across farm, peri-urban, and wetland habitats. We used three trap types at three different latitudes across Europe (Sweden, the Netherlands, Italy). Next to classical community parameters (diversity, richness, and evenness), we used non-metric multidimensional scaling (NMDS) analyses to assess community ecology statistics. Higher mosquito species diversity was observed from collections with the Mosquito Magnet Liberty Plus compared to the BG Sentinel trap. Highest mosquito species diversity was found in Sweden. Within Sweden, mosquito diversity was highest in wetland habitats, whereas it was highest at farms for the Netherlands and Italy. In the Netherlands most midge species were collected, whereas midge species diversity was highest in Italy when compared to Sweden and the Netherlands. NMDS analyses further revealed that community composition was distinctly different between countries for both mosquitoes and midges. Differences in habitat communities were only found within countries. Community patterns seen in mosquitoes were dissimilar from those in midge species. Different vector species diversity might be caused by a larger variety in breeding places, optimal (micro)climate or availability of host species. Lower diversity in vector communities, thus higher abundance of a few dominant species could increase the risk of vector-borne diseases. Given the fact that many vector-borne diseases require a set of multiple species that together influence the rate of transmission, understanding the ecology of vector networks is becoming increasingly important.

Host preference of mosquitoes mediated by skin bacterial volatiles

N.O. Verhulst[1], A. Busula[2] and W. Takken[1]
[1]Wageningen University, Laboratory of Entomology, Droevendaalsesteeg 1, 6708 PB Wageningen, the Netherlands, [2]International Centre of Insect Physiology and Ecology, P.O. Box 30772-00100 GPO, Nairobi, Kenya; niels.verhulst@wur.nl

Skin microbes are important in human health and disease. In vector-borne diseases the skin microbiota mediate the interaction between vertebrates and arthropod vectors. We have shown that odorants from the microbiota on the human skin guide anthropophilic malaria mosquitoes towards their blood-meal hosts. In this study we tested the attractiveness of natural and synthetic odour blends to mosquitoes with different host preferences and their response to volatiles released from skin bacteria. Different odour baits elicited varying responses among mosquito species in both semi-field and field setup. Synthetic odour blends were highly effective for trapping mosquitoes; however, not all mosquitoes responded equally to the same odour blend. Volatiles from skin bacteria collected from different hosts and grown on agar plates induced differential responses of the anthropophilic *Anopheles gambiae s.s.* mosquito and more zoophilic *An. arabiensis*, which matched their response to the host odours themselves in the first series of experiments. Interestingly, *An. gambiae* had a specialized response to volatiles from four specific bacteria, common on the human skin, while *An. arabiensis* responded equally to all bacterial species tested. Skin bacterial volatiles may play an important role in guiding mosquitoes with different host preferences to their specific host. Identification of these bacterial volatiles can contribute to development of new synthetic odour blends that may be used for sampling of mosquitoes with different host preferences.

Understanding the role of host emitted semiochemicals on *Culicoides* (Diptera: Ceratopogonidae) behaviour

M.A. Miranda[1], C. Barceló[1], D. Borràs[1] and A.C. Gerry[2]
[1]University of the Balearic Islands, Biology, Cra. Valldemossa km 7.5, 07122 Palma Mallorca, Spain, [2]University of California, Riverside, Riverside, USA; ma.miranda@uib.es

Biting midges of the genus *Culicoides* (Diptera: Ceratopogonidae) include several species which are important vectors of arboviral diseases such as bluetongue and African horse sickness. Females are haematophagous and usually feed on different species of animal host, which are located in the environment by detecting chemical cues (kairomones) emitted by them. Interestingly, and according to literature, *Culicoides* respond differently to generalist chemical cues such as octenol or lactic acid when compared to other haematophagous insects such as mosquitoes. In order to explore the effect of different semiochemicals on trapping of *Culicoides* adults, we carried out two trials in a mixed animal farm in Majorca (Balearic Islands, Spain) in June-July 2016 using acetone, octenol, lactic acid and ammonia and a combination of the same attractants plus dry ice as a source of CO_2. All attractants were tested in vials with a constant rate of odour release placed on Onderstepoort traps with the UV light removed. In addition, two UV light Onderstepoort traps were used as reference of the presence of *Culicoides* adults in the area. The trial was conducted during 8 nights in total and the attractants were rotated clockwise each night. General results showed a poor performance of captures of *Culicoides* in all attractants compare to the Onderstepoort light traps. Species captured during this study included Obsoletus group species, *C. imicola*, *C. newsteadi*, *C. circumscriptus*, as well at other *Culicoides* spp. in small numbers. The odors, both alone or with CO_2, did not increased the collection of these *Culicoides* spp. relative to a control. We discuss the chemical ecology of *Culicoides* spp. in relation to the obtained results and the necessity of improving trapping techniques based on specific attractants.

Species composition, activity patterns and blood meal analysis of sand fly populations in the metropolitan region of Thessaloniki

A. Chaskopoulou[1], I.A. Giantsis[1], S. Demir[2] and M.C. Bon[3]
[1]USDA-ARS, European Biological Control Laboratory, Tsimiski 43, 54623, Greece, [2]Ege University, Department of Biology, Bornova, Izmir, 35100, Turkey, [3]USDA-ARS, European Biological Control Laboratory, Campus International de Baillarguet, Montferrier-sur-Lez, 34988, France; achaskopoulou@ars-ebcl.org

Species composition, activity patterns and blood meal analysis of sand fly populations were investigated in the metropolitan region of Thessaloniki, North Greece from May to October 2011. Sampling was conducted weekly in 3 different environments (animal facilities, open fields, residential areas) along the outskirts of the city in areas of increased canine leishmania transmission. Six sand fly species (*Phlebotomus perfiliewi, Phlebotomus tobbi, Phlebotomus simici, Plebotomus papatasi, Sergentomya minuta* and *Sergentomya dentata)* were identified using both classical and molecular techniques. DNA barcodes were characterized for the first time for two (*P. simici* and *S. dentata)* of the six recorded species. Phylogenetic analysis based on the COI gene sequences confirmed the grouping of *P. tobbi, P. perniciosus* and *P. perfiliewi* (subgenus *Larrousius)* and the monophyly of *P. simici* (subgenus *Adlerius)*. By far the most prevalent species was *P. perfiliewi*, followed by *P. simici* and *P. tobbi*. The largest populations of sand flies were collected from animal facilities, followed by residential areas and open agricultural fields. Peak activity of sand flies overall occurred mid-August to mid-September and then declined sharply in October. Blood meal analysis showed that *P. perfiliewi* and *P. simici* feed preferentially on humans (88% and 95%, respectively) but also feed on chickens and goats. When designing a control strategy to alleviate sand fly nuisance in the region of Thessaloniki the following conclusions can be reached from this study: (a) August and September are high risk months due to increased sand fly activity levels; (b) animal facilities within or adjacent to urban settlements are high risk areas and may act as a maintenance and amplification foci for the vector as well as the parasite; and (c) the abundance, ubiquity and feeding behavior of *P. perfiliewi* and *P. simici* establishes them as potentially important vectors of *Leishmania* in the region.

Prediction of mosquito occurrence patterns in urban area using machine learning algorithms

D. Lee and Y.-S. Park
Kyung Hee University, Dongaemun, Seoul 02447, Republic of Korea; parkys@khu.ac.kr

Mosquito-borne infectious diseases are a major public health problem in the world. Increase of mosquito populations and expansion of distributions are highly influenced by their habitat condition. Climate and meteorological variables are commonly used for predicting the occurrence of mosquito population and severity of mosquito-borne epidemics. In this study, we predicted the occurrence patterns of mosquitos in urban area. We used the mosquito occurrence data collected using CO_2-baited traps at the sampling sites. Three meteorological variables including temperature, humidity and precipitation were measured concurrently at the nearest automatic weather station. Multivariate analyses (cluster analysis and principal component analysis; PCA) were used to characterize land use patterns. The monitoring sites were differentiated into three clusters based on differences in land use types such as culture and sport areas, inland water, residential, and commercial areas. These clusters were well reflected in PCA ordinations, indicating the land use effect on mosquito occurrence. Two classification models, classification and regression tree (CART) and random forest (RF), were applied to determine the relationships with urban mosquito occurrence and meteorological variables, and also used as prediction models for two different land use groups. The usage of the classification models can guarantee the prediction of the mosquito occurrence based on the land use types and meteorological factors.

Species composition of potential arbovirus vector mosquitoes in selected areas of South Africa

T. Johnson[1], T. Nelufule[1], A.P.G. Almeida[1,2] and L. Braack[1,3]
[1]Centre for Viral Zoonoses, Department of Medical Virology, University of Pretoria, Pathology Building, 5 Bophelo Road Prinshof Campus, University of Pretoria, Private Bag X20, Hatfield 0028, South Africa, South Africa, [2]Global Health and Tropical Medicine, Instituto de Higiene e Medicina Tropical, Universidade NOVA de Lisboa, R Junqueira 100, 1349-008 Lisboa, Portugal, [3]Institute for Sustainable Malaria Control, University of Pretoria, Prinshof Campus, University of Pretoria, Private Bag X20, Hatfield 0028, South Africa, South Africa; palmeida@ihmt.unl.pt

A survey was conducted in four sampling sites, two in Gauteng (Kyalami and Boschkop), and two in Limpopo (Marakele and Lapalala) Provinces within South Africa, with the objective of sampling mosquito vectors of zoonotic arboviruses. Mosquito tent traps (MTT) and CDC light traps (CDC) baited with carbondioxide were used to gather information on adult mosquito species composition, distribution and seasonality. A total of 5,543 mosquitoes (Diptera: Culicidae) representing 35 species from 9 genera: *Aedomyia*, *Aedes*, *Anopheles*, *Coquillettidia*, *Culex*, *Ficalbia*, *Mimomyia*, *Mansonia* and *Uranotaenia* were collected between May 2015 and May 2016. The most widely distributed potential arbovirus vector species collected during this study were *Culex theileri* and *Culex univittatus*. These two species were also the most abundant in Gauteng Province. *Culex poicilipes* was, however, the most abundant at Marakele National Park and Lapalala Wilderness Game Reserve. Of the aedine mosquitoes, *Aedes vittatus*, dominated in both trap types at Lapalala during the months of September 2015 to April 2016. This mosquito is a known vector of yellow fever virus, and has recently been reported as a potential vector of chikungunya virus. It is also evident from the study that the distribution and species composition of these mosquitoes varies with site and season. However, of particular importance in terms of arbovirus transmission in South Africa are the two species: *Culex theileri* and *Culex univittatus* which are known to be able to transmit Rift Valley fever, West Nile fever and sindbis viruses in South Africa and elsewhere.

Does artimisinin based combination therapy influence mosquito fitness and host-seeking behaviour?

J.G. De Boer, A.O. Busula, J. Ten Berge, T.S. Van Dijk and W. Takken
Wageningen University, Laboratory of Entomology, Droevendaalsesteeg 1, 6708 PB Wageningen, the Netherlands; jetske.deboer@wur.nl

Artemisinin-based combination therapy (ACT) is recommended against malaria in many endemic areas, and thus widely used. Surprisingly little is known about the effect of ACTs on mosquitoes that transmit malaria parasites. Our objectives were to: (1) determine whether ACTs have a direct impact on mosquito fitness when added to the bloodmeal; and (2) evaluate whether ACTs influence mosquito host-seeking behaviour by changing the intrinsic attractiveness of human skin odour. Our study was done with *Anopheles gambiae s.l.*, which is the main vector of malaria in sub-Saharan Africa. Mosquitoes that were fed on blood with ACT survived equally long as mosquitoes fed on control blood. ACT-fed and control mosquitoes also laid equal numbers of eggs, thus fitness was not affected by the treatment. To investigate host-seeking behaviour of mosquitoes, adult malaria-free men were given a treatment dose of an ACT, and skin odour was collected on nylon socks before, during and three weeks after treatment. A screenhouse choice test showed no preference of *An. gambiae* females between socks worn by the same person before, during or after ACT-treatment. Relative attractiveness of nylon socks to *An. coluzzii* in a dual-choice olfactometer was also not influenced by ACT-treatment although mosquitoes appeared to be more responsive to skin odour collected three weeks after ACT-treatment. We conclude that ACT-treatment does not affect fitness and host-seeking behaviour of malaria mosquitoes. Our results are important in light of possible transmission of gametocytes from ACT-treated people to malaria vectors.

Annual population dynamics of hematophagous dipterans in Israeli dairy farms suggest a potential vector of lumpy skin disease

Y. Gottlieb, E. Kahana and E. Klement
The Hebrew University, Koret School of Veterinary Medicine, POB 12, Rehovot, 76100, Israel; yuvalgd@yahoo.com

Lumpy skin disease (LSD), a viral disease affecting Bovidae, recently invaded into the Middle-East and also Europe. The LSD virus (LSDV) is considered to be mechanically transmitted by arthropod vectors. However, it is not known which arthropod can transmit the virus under field conditions. In order to determine the possible vector of LSDV, we performed a yearlong trapping of dipterans once a month during 2014 in dairy herds which were affected in an LSD epidemic that occurred during 2012-2013 in Israel. For each farm, the monthly relative abundance of each dipteran (including mosquitoes, biting midges, filth flies and sand flies) was calculated by dividing the number of dipteran caught during each month by the annual number of the same dipteran, caught along the entire year. During the month parallel to the month of outbreak onset the relative abundance of *Stomoxys calcitrans*, the stable fly, was significantly higher than the relative abundance of other hematophagous dipterans. The highest abundance of the stable fly was observed during December and April, the months in which most dairy cattle herds were affected in 2012-2013. A model based on land surface temperature, ambient temperature and rainfall was fitted to the 2014 data and showed a significant correlation with the actual relative abundance of the stable fly during 2014 (R^2=0.82). This model was then used to estimate the stable fly relative abundance during 2012-2013 and found similar correlation between its abundance and LSD outbreak onset in the dairy farms in the study area. A significantly lower abundance of the stable fly was observed during October and November, despite the existence of disease in adjacent grazing beef herds. We therefore conclude that of the stable fly is the primary suspect for being the vector of LSD in zero-grazing dairy cattle, and that another vector may be involved in the transmission of the disease in grazing beef herds. These results should be followed by laboratory studies assessing the vector competence of *S. calcitrans* for transmission of LSD virus.

Trap yield and species composition comparison for arbovirus vector mosquito surveillance in South Africa

A.P.G. Almeida[1,2], T. Nelufule[1], R. Swanepoel[1] and L. Braack[1]
[1]Centre for Viral Zoonoses, Department of Medical Virology, University of Pretoria, Private Bag X20, Hatfield 0028, South Africa, [2]Global Health and Tropical Medicine, Instituto de Higiene e Medicina Tropical, Universidade NOVA de Lisboa, R Junqueira 100, 1349-008 Lisboa, Portugal; palmeida@ihmt.unl.pt

Arbovirus vector mosquito surveillance programs rely on efficient trapping. Mosquito infection rates in nature are usually quite low, hence surveillance program success relies on adequate sample sizes, i.e. large samples. Different kinds of traps are used, most often Centers for Disease Control (CDC) miniature light trap variants or mosquito tent traps (MTT). A survey was conducted at six sampling sites in three Provinces within South Africa; two in Gauteng (Kyalami and Boschkop), three in Limpopo (Marakele and Lapalala, Shingwedzi), and one in Free State (Hertzogville), with the objective of comparing trap yield and species composition between different trap types. MTT, CDC incandescent light traps (CDC-Inc), CDC white ultra violet light (CDC-W), and CDC black ultra violet light (CDC-B), all baited with carbon dioxide (dry ice), were used simultaneously at the same sites, placed *ca.* 50 m apart from each other, between January and June 2014. A total of 220 collections were performed, 55 of each trap type. The average number of mosquitoes per trap was 36±59 (SD) (0-369) for MTT, significantly less than any of the CDC trap types, which yielded 72±96 (0-349) for CDC-B, 67±99 (0-545) for CDC-W, and 55±76 (SD) (0-435) for CDC-Inc. However, the various CDC trap yields were not statistically significantly different from each other. As to the species composition among trap types, mosquito genera were differently collected by the various trap types, namely, those belonging to the genera *Aedes* (whether subgenus *Stegomyia* or other, were higher in CDC-B), *Coquillettidia* (higher in CDC-W/Inc), or *Culex*, though without clear differences between various CDC traps but higher than MTT, whereas *Anopheles* and *Mansonia* were not different among all. However, it should also be noted that the distribution of these various mosquito genera were all differently distributed across these six surveyed sites.

Study of the bionomics of the *Obsoletus* complex and other livestock associated *Culicoides* at different temperatures in laboratory conditions

C. Barceló and M.A. Miranda
University of the Balearic Islands, Biology. Laboratory of Zoology, Cra. Valldemossa km 7.5, 07122, Spain; carlos.barcelo@uib.es

Females of several biting midges species from genus *Culicoides* (Diptera: Ceratopogonidae) transmit important arboviruses such as bluetongue, African horse sickness, Schmallenberg and epizootic haemorrhagic disease viruses. The basic bionomics of the major vector and non-vector *Culicoides* species associated to farms remains mostly unexplored. In this work we studied the *Culicoides* sub-adult development at 3 different temperatures in laboratory conditions of the vector species *C. obsoletus* as well as the non-vector species *C. circumscriptus*, *C. paolae* and *C. cataneii*. Insects were collected from field between spring and autumn 2015 in two cattle farms located in Mallorca (Balearic Islands, Spain). Gravid females were kept individually at 15, 25 and 30 °C in cardboard boxes with moistened cotton wool and filter paper as a substrate for oviposition. Eggs were transferred to Petri dishes with 2% agar gel medium and larvae were reared in the agar containing the nematode *Panagrellus redivivus*. Pupae were transferred again to cardboard boxes till adults emerge. We present results of bionomic parameters for the different *Culicoides* species, such as number of eggs laid per females, percentage of hatch, time for development from larva to pupa, percentage of pupation and adult survival.

Shift in species composition in the *Anopheles gambiae* complex after implementation of longlasting insecticidal nets in Dielmo, Senegal

S. Sougoufara[1,2], M. Harry[3], S. Doucoure[2], P.M. Sembene[1] and C. Sokhna[2]
[1]Universite Cheikh Anta Diop Dakar, Biologie Animale, Fann, Senegal, 5005 Dakar, Senegal, [2]URMITE, UMR sur les Maladies Infectieuses et Tropicales Emergentes, Routes des Peres Maristes, 1386 CP 18524 Dakar, Senegal, Senegal, [3]UMR Évolution, Génomes, Comportement, Écologie (EGCE), CNRS-IRD Université Paris Sud, Gif-sur-Yvette, France, 91198 Gif-sur-Yvette, Paris, France; seynabou.sougoufara@gmail.com

Over 90% of deaths from malaria have occurred in sub-Saharan Africa. The last decade has seen impressive progress in malaria control in this area, primarily because of several concurrent control initiatives, particularly antivectorial programmes with the use of long-lasting insecticidal nets (LLINs) and indoor residual spraying (IRS) more heavily. However, the effectiveness of these control tools depends on vector ecology and behaviour, which also largely determine the efficacy of certain *Anopheles* mosquitoes as vectors. Malaria vectors in sub-Saharan Africa are primarily species of the *Anopheles gambiae* complex, which present intraspecific differences in behaviour that affect how they respond to vector control tools. The focus of this study is the change in species composition in the *An. gambiae* complex after the implementation of LLINs in Dielmo, Senegal. The main findings showed just after the implementation of LLINs in 2008, 82.91% of individuals sampled were *An. arabiensis*, up from just 21.43% in 2006. By 2010, the proportion of collections represented by *An. arabiensis* had decreased to 60.00%. *Anopheles coluzzii* was the most prevalent species in 2006, before the implementation of LLINs (57.14%); however, its density decreased gradually from 2008 (13.67% of collections), when LLINs were first implemented in Dielmo, to 2010 (8.44% of collections). The frequency of *An. gambiae* decreased between 2006 (19.23% of collections) and 2008 (2.99% of collections) and increased to a much higher density in 2010 (31.11% of collections).This observation does not mean that vector control tools have failed because these tools may have dramatically reduced the densities of those vectors that are most often in contact with human hosts. However, a higher usage of bednets must be promoted in Dielmo combined with additional vector control tools that can target the full range of malaria vectors.

Assessing male *Anopheles gambiae, Anopheles coluzzii* and their reciprocal hybrids swarming behaviour in contained semi-field

C. Nignan[1,2], A. Niang[2], H. Maïga[2], S.P. Sawadogo[2], K.R. Dabiré[2] and A. Diabaté[2]
[1]Master International d'Entomologie (MIE), Université de Montpelier et CEMV de Bouaké, Faculté des Sciences, France, [2]Institut de Recherche en Sciences de la Santé (IRSS), Entomologie medicale, Bobo-Dioulasso/Burkina Faso, 01 BP 545 Bobo-Dioulasso 01, Burkina Faso, Burkina Faso; nignancharles@yahoo.fr

An. gambiae and *An. coluzzii* are two of the most important malaria vector species in sub-Saharan Africa. These recently-diverged sibling species do not exhibit intrinsic pre-zygotic barriers to reproduction and are thought to be separated by strong assortative mating combined with selection against hybrids. The main objective of this study was to assess the capacity of *An. gambiae, An. coluzzii* and their reciprocal hybrids to swarm above artificial visual markers in the malaria sphere. Four cages were prepared in the malaria sphere. In each cage were placed three swarm markers (blacksheet, woodpile and bare ground). Male mosquitoes of the parental species and their reciprocal hybrids were generated by forced mating technics. A standard number of sexually active males from the cross mating called hereafter *An. coluzzii* (400), Hibrids col/gam and gam/col (400) and *An. gambiae* (400), respectively were released separately in three control cages. In addition a mix of 100 males of each of the four groups differently colored was released in a same room, as a test cage. The number of swarming males were counted and compared between groups. Our results showed a significant difference in parental species success at swarming compared to that of hybrids (P=0.022) both in control and test cages. No swarming hybrids were collected out of 1,600 males that have been released in the control cage. Furthermore, the number of male *An. coluzzii* collected above black sheet (60.4%) and above wood pile (6.67%) were significantly higher than those of males collected above bare ground, though more male *An. gambiae* were swarming above bare ground in the test cage. This study would contribute to better understand the process of speciation between *An. gambiae* and *An. coluzzii* and could help in guiding future vector control engineering programs against malaria.

The effect of interspecific competition between *Culex pipiens* and *Aedes albopictus* in northern Italy

G. Marini[1,2], G. Guzzetta[3], F. Baldacchino[1], D. Arnoldi[1], F. Montarsi[4], G. Capelli[4], A. Rizzoli[1], S. Merler[3] and R. Rosà[1]
[1]Fondazione Edmund Mach, Via Mach 1, 38010 San Michele all'Adige, Italy, [2]University of Trento, Via Sommarive 14, 38123 Trento, Italy, [3]Fondazione Bruno Kessler, Via Sommarive 18, 38123 Trento, Italy, [4]Istituto Zooprofilattico Sperimentale delle Venezie, viale dell'Università 10, 35020 Legnaro, Italy; giovanni.marini@fmach.it

Aedes albopictus and *Culex pipiens* larvae reared in the same breeding site compete for resources, with an asymmetrical outcome that disadvantages only the latter species. The impact of these interactions on the overall ecology of these two mosquito species has not yet been assessed in the natural environment. In the present study, the temporal patterns of adult female mosquitoes from both species were analyzed in northeastern Italy, and substantial temporal shifts between abundance curves of *Cx. pipiens* and *Ae. albopictus* were observed in several sites. To identify the drivers of such differences, we developed a density-dependent mechanistic model that takes explicitly into account the effect of temperature on the development and survival of both species. In addition, we included into the model the effect of asymmetric interspecific competition, by adding a mortality term for *Cx. pipiens* larvae proportional to the larval abundance of *Ae. albopictus.* A model calibration was performed through a Monte Carlo Markov chain approach using capture data collected in our study sites in 2014 and 2015. In several cases, our results show that an increasing in abundance of *Ae. albopictus* caused an early decline of *Cx. pipiens* population and that the competition effect was enhanced by higher abundance of either species. However, in some cases temporal shifts can also be explained in the absence of interspecific competition resulting from a 'temporal niche' effect, when the optimal fitness environmental conditions for the two species are reached at different times of the year. These findings demonstrate the importance of taking into account ecological interactions between mosquito species in temperate climates, with important implications for the invasion dynamics of the alien species and for risk assessment of mosquito transmitted pathogens.

Diversity and role of flying and aquatic insects in the ecology of *Mycobacterium ulcerans* in Benin, West Africa

B. Zogo, E. Marion, J. Babonneau, C. Pennetier, M. Akogbeto, L. Marsollier and A. Djenontin
Master International d'Entomologie (MIE)/Centre de Recherche Entomologique de Cotonou (CREC), Ministère de la Santé, BP70 Allada, Benin; mahzb2000@yahoo.fr

Buruli ulcer (BU) is a necrotizing skin disease caused by the environmental bacteria, *Mycobacterium ulcerans* (M.u.). The mode of transmission of this emerging neglected disease remains an enigma although various studies from endemic areas of West Africa and Australia attempted to incriminate water bugs and mosquitoes as vectors respectively. The objective of this study was to investigate the diversity of flying and aquatic insects and their involvement in the ecology of M.u. in high BU endemic areas of Benin, West Africa. Flying and aquatic insects were collected from two BU endemic villages in the department of Oueme in southeast Benin using CDC light traps and square nets respectively. The presence of M.u. DNA in the insect samples was detected by quantitative PCR. A total of 1,556 flying insects were sampled into 7 taxonomic orders of which 480 mosquitoes distributed among 9 species. As well, 1,221 aquatic bugs belonging to 7 families, 303 Coleoptera and 220 mosquito larvae were collected. No trace of the bacteria was found in mosquitoes and other flying insects while aquatic insects including water bugs were found positive to M.u. DNA. Our study provides strong support for the role of water bugs as hosts and likely as vectors of M.u. On the contrary, it reveals that flying insects including mosquitoes do not play a pivotal role in the ecology of M.u in our selected endemic areas.

Evidence that agricultural pesticides select for insecticide resistance in the malaria vector *Anopheles gambiae*

S. Djogbenou, B. Assogba, L. Djossou and M. Makoutode
Institut Régional de Santé Publique/Université d'Abomey-Calavi, Santé Environnement, Route des Exclaves, Ouidah, BP 384 Ouidah, Benin; ldjogbenou22002@yahoo.fr

The spread of pyrethroid resistance in malaria vectors is of concern as there are no alternative insecticides approved for use on insecticide treated nets. Although much of the current resistance can be attributed to massive scaling up of malaria control interventions, the role of insecticide use in agriculture cannot be ignored. The aim of the present study was to investigate the possible relation between the agricultural use of insecticides and the emergence and the evolution of insecticide resistance in *Anopheles gambiae*. Bioassays were conducted using simulated mosquito larval habitats to evaluate the effect of soil samples history trait of *Anopheles gambiae* with know insecticide resistance mechanisms. Soil samples were collected from vegetable farms areas of Houeyiho in Benin, including, site with recent insecticide use, site where insecticides had not been used for two months, and where insecticides had not been used (control). Pupation and emergence rates were very low in pyrethroid-susceptible strains when exposed to soil that had been recently treated. Pupation and emergence rates in strains with the kdr mutation were much higher, around 60% in strains with both the kdr and ace-1 mutations and around 70% when kdr alone was present. The fact that strains with the kdr mutation survived at higher rates than strains without kdr is consistent with the reported use of pyrethroids at the site. This effect was not observed when soil from areas which was not treated in the previous two months, indicating degradation of the insecticide. Although this study is observational and does not take differences in soil composition into account, use of pyrethroids is expected to play a role in the emergence and the evolution of insecticide resistance in *Anopheles*. This aspect should be taken into account to better use the insecticide in the context of integrated pest management programs.

Blood influence on protein spectra of sand fly females and host blood identification using MALDI-TOF mass spectrometry

K. Hlavackova[1], V. Dvorak[1], P. Halada[2] and P. Volf[1]
[1]Faculty of Science, Charles University in Prague, Department of Parasitology, Albertov 6, 128 43, Praha, Czech Republic, [2]Institute of Microbiology, Academy of Science, Videnska 1083, 142 20, Praha, Czech Republic; hlavackova.k@centrum.cz

Protein profiling by matrix-assisted laser desorption/ionization time-of-flight (MALDI-TOF) mass spectrometry has proved to be a suitable method for species identification of vectors, including sand flies. Females of the genera *Phlebotomus* and *Lutzomyia* stay the only proven vectors of leishmaniasis, and identification of their blood meals sources is crucial for better understanding host/vector interactions. It could also help with setting a proper control over these vectors in endemic areas. Nowadays, the methods used for host-blood identification are time consuming and costly, therefore mass spectrometry would be a solution since this approach is fast and with minimal sample preparation. We studied influence of blood on the quality of protein profiles of females in different time points during the whole gonotrophic cycle, which includes the influence of developing eggs as well. We also tested the possibility of using protein profiling for blood meal identification in sand fly females experimentally fed on different vertebrate hosts. These female specimens were also analyzed at various points in time and protein profiles of abdomen containing host blood were compared with protein spectra obtained from host blood only. Sand fly females were fed either on one or more hosts. In conclusion, MALDI-TOF protein profiling seems to be a useful method not only for species identification of many organisms, but also a helpful approach for identifying blood meal in engorged females.

Host-feeding patterns of mosquito species in Germany

J. Börstler[1], H. Jöst[1,2], R. Garms[1], A. Krüger[3], E. Tannich[1,2], N. Becker[4,5], J. Schmidt-Chanasit[1,2] and R. Lühken[1]
[1]Bernhard Nocht Institute for Tropical Medicine, Bernhard-Nocht-Straße 74, 20359 Hamburg, Germany, [2]German Centre for Infection Research (DZIF), partner site Hamburg-Luebeck-Borstel, Inhoffenstraße 7, 38124 Braunschweig, Germany, [3]Department of Tropical Medicine, Bundeswehr Hospital Hamburg, Bernhard-Nocht-Straße 74, 20359 Hamburg, Germany, [4]University of Heidelberg, Grabengasse 1, 69117 Heidelberg, Germany, [5]German Mosquito Control Association (KABS), Institute for Dipterology, Georg-Peter-Süß-Str. 3, 67346 Speyer, Germany; renkeluehken@gmail.com

Mosquito-borne pathogens are of growing importance in many countries of Europe including Germany. At the same time, the transmission cycles of most mosquito-borne pathogens are not completely understood. There is especially a lack of knowledge about the vector capacity of the different mosquito species, which is strongly influenced by their host-feeding patterns. From 2012 to 2015, 775 blood-fed mosquito specimens were collected and analysed for their hosts. The host species for each mosquito specimen was identified with polymerase chain reactions and subsequent Sanger sequencing of the cytochrome *b* gene. A total of 32 host species were identified for 23 mosquito species, covering 21 mammalian species (including humans) and eleven bird species. Although the collected blood-fed mosquito species had a strong overlap of host species, two different host-feeding groups were identified with mosquito species feeding on (1) non-human mammals and humans; or (2) birds, non-human mammals, and humans, which make them potential vectors of zoonotic pathogens only between mammals or between mammals and birds, respectively. The presented study indicates a much broader host range compared to the classifications found in the literature, which highlights the need for studies on the host-feeding patterns of mosquitoes to further assess their vector capacity and the ecology of zoonoses in Europe.

The hitch-hiking tiger: first experimental study on the passive transportation of *Aedes albopictus* by cars

R. Eritja[1], D. Roiz[2], J.R.B. Palmer[3], I. Sanpera-Calbet[1] and F. Bartumeus[3]
[1]Consell Comarcal del Baix Llobregat, Servei de Control de Mosquits, Parc Torreblanca, 08980 Sant Feliu de Llobregat, Spain, [2]IRD, MIVEGEC, 911 Avenue Agropolis, 34394 Montpellier Cedex 05, France, [3]CSIC, CEAB, Accés Cala Sant Francesc 14, 17300 Blanes, Girona, Spain; reritja@elbaixllobregat.cat

Whereas the invasive vector mosquito *Aedes albopictus* (Diptera: Culicidae) has low active dispersal abilities, its worldwide colonization process is being extremely fast. Substantial indirect evidence suggests a major role of passive transportation along roads due to individual mosquitoes entering and later exiting vehicles operated by humans. However, this has never been experimentally checked because of the difficulties associated with real-time roadside surveys. This experimental study on the presence of *Aedes albopictus* inside circulating cars was performed in collaboration with the Catalan Road Police Department. After an initial calibration test to assess the sampling efficiency, a total of 358 cars were stopped and vacuumed in routine police controls in motorways, and 412 more vehicles were sampled at Technical Inspection Stations (ITV) in the Baix Llobregat area, near Barcelona city. Drivers were interviewed for basic mobility information. For the first time, we obtained experimental evidence of the role of passive transportation in vehicles as a major driving force for *Ae. albopictus* dispersion in Europe. Results are presented and discussed in the context of the current species invasion in Spain.

Chemosensory responses to the repellent *Nepeta* oil and its major component nepetalactone by the yellow fever mosquito, a vector of zika virus

J.C. Dickens, J.T. Sparks and J.D. Bohbot
USDA, ARS, NEA, BARC, Invasive Insect Biocontrol and Behavior Laboratory, 10300 Baltimore Avenue, Beltsville, MD 20705, USA; joseph.dickens@ars.usda.gov

Nepeta oil (catnip) and its major component, nepetalactone, have long been known to repel insects including mosquitoes. However, the neural mechanisms through which these repellents are detected by mosquitoes, including the yellow fever mosquito, *Aedes aegypti*, an important vector of zika virus, were poorly understood. Here we show that *Nepeta* oil volatiles activate olfactory receptor neurons within the basiconic sensilla on the maxillary palps of female *Ae. aegypti.* A gustatory receptor neuron sensitive to the feeding deterrent quinine and housed within sensilla on the labella of females was activated by both *Nepeta* oil and nepetalactone. Activity of a second gustatory receptor neuron sensitive to the feeding stimulant sucrose was suppressed by both repellents. Our results provide neural pathways for the reported spatial repellency and feeding deterrence of these repellents. A better understanding of the neural input through which female mosquitoes make decisions to feed will facilitate design of new repellents and management strategies involving the use of repellents.

Monitoring and control of a large population of *Aedes albopictus* in an urban area of Southwest Germany

A. Jöst[1], I. Ferstl[1,2], B. Pluskota[1], S. Schön[1,2], E. Tannich[3], C. Kuhn[4], A. Mosca[5], M. Giovannozzi[5] and N. Becker[1]
[1]Institute for Dipterology/Kommunale Aktionsgemeinschaft zur Bekämpfung der Schnakenplage (KABS), Georg-Peter-Süß-Straße 3, 67346 Speyer, Germany, [2]University of Freiburg, Friedrichstr. 39, 79098 Freiburg, Germany, [3]Bernhard Nocht Institute for Tropical Medicine, Bernhard-Nocht-Straße 74, 20359 Hamburg, Germany, [4]German Environment Agency, Bötticher Str. 2, 14195 Berlin, Germany, [5]Piedmont Mosquito Abatement District-Istituto per le Piante da Legno e l'Ambiente (IPLA), Corso Casale 476, 10124 Torino, Italy; artur.joest@kabs-gfs.de

In July 2015, a large population of *Aedes albopictus* was detected in the city of Freiburg in Southwest Germany, following the lead given by an attentive citizen. The distribution focus was found to be located on an allotment garden, in close proximity to a train station terminal, which is part of a so called 'Rolling Highway (ROLA)' and which was likely the route of mosquito introduction. The ROLA is a transport system that carries complete trucks (cabs and trailers) by rail. All trucks arriving in Freiburg come from Italy, the country with the highest population density of *Aedes albopictus* in Europe. Immediately after discovery of the mosquitoes, comprehensive monitoring and control measures were carried out in order to reduce population density and to get an overview on the distribution in the Freiburg area. The results indicated that the population was largely restricted to the allotment garden although a moderate spread of *Aedes albopictus* was observed within a radius of 600 m from the hotspot. In total more than 4,000 Asian tiger mosquitoes could be caught on the site and in the vicinity of the allotment garden. This illustrates the enormous size of the population, which prompted us to implement immediate control measures. In addition, in spring 2016, containers of the allotment garden were thoroughly cleaned to reduce the number of possible diapausing eggs. The abrasions of the containers were collected and flooded in the laboratory. In 21% of the samples larvae hatched and thus provided evidence for successful overwintering on the allotment garden. After occurrence of the first larvae in the field, intensive monitoring and control measurements started again. Results will be presented.

High-resolution density map of the tick-borne encephalitis and Lyme borreliosis vector *Ixodes ricinus* for Germany

K. Brugger and F. Rubel
University of Veterinary Medicine Vienna, Institute for Veterinary Public Health, Veterinaerplatz 1, 1210 Vienna, Austria; katharina.brugger@vetmeduni.ac.at

The castor bean tick *Ixodes ricinus* is the principal vector for a variety of viral, bacterial, and protozoan pathogens causing a growing public-health issue over the past decades. Density maps of *I. ricinus* are needed to quantify the risk of tick-borne diseases, but they are still rare especially on landscape or regional scales. Here, a high-resolution map of questing *I. ricinus* nymphs was compiled for Germany covering an area of about 357,000 km^2 (regional scale). Input data comprise mean annually accumulated nymphal density, as observed by monthly flagging an area of 100 m^2 collected at 69 sampling sites between 2006 and 2014. A negative binomial regression model was developed to interpolate the observed tick densities to unsampled locations by using bioclimatic variables and land cover. The variables were selected according to their significance by the Akaike Information Criterion (AIC). The tick map was verified by leave-one-out cross-validation. The root mean square error of RMSE = 126 nymphs per 100 m^2 is of the order of the inter-annual variation of the tick densities. Tick densities are very low in urban (green) areas. Maximum annual densities up to 1,000 nymphs per 100 m^2 are observed in broad-leaved forests, mainly in the southern federal states as Bavaria or Baden-Württemberg. The compilation of a high-resolution density map of unfed nymphal *I. ricinus* for Germany provides a novel, nationwide insight into the distribution of an important disease vector.

Impacts of introgressive hybridisation in the population structure and bionomics of *Anopheles gambiae* in Guinea Bissau

J.L. Vicente[1], C.S. Clarkson[2], B. Caputo[3], B. Gomes[1], M. Pombi[3], C.A. Sousa[1], T. Antao[2], G. Botta[3,4], E. Mancini[3], V. Petrarca[3], D. Mead[5], E. Drury[5], J. Stalker[5], A. Miles[4], D.P. Kwiatkowski[4,5], M.J. Donnelly[2], A. Rodrigues[6], A. Della Torre[3], D. Weetman[2] and J. Pinto[1]
[1]Instituto de Higiene e Medicina Tropical, Universidade Nova de Lisboa, Global Health and Tropical Medicine, 1349-008 Lisbon, Portugal, [2]Liverpool School of Tropical Medicine, Liverpool L3 5QA, United Kingdom, [3]Università di Roma Sapienza, Roma, 00185, Italy, [4]Wellcome Trust Centre for Human Genetics, Oxford, SW6 7PP, United Kingdom, [5]Wellcome Trust Sanger Institute, Hinxton CB101SA, United Kingdom, [6]Instituto Nacional de Saude Publica, Bissau, 861-1004, Guinea-Bissau; jpinto@ihmt.unl.pt

Anopheles coluzzii and *Anopheles gambiae* are characterized by low levels of hybridization along their sympatric range in West Africa. However, high hybridization with asymmetric introgression has been found in the west African limit of their distribution, particularly in Guinea Bissau. In this study, we have assessed genetic and genomic variation and analyzed traits of medical importance in *An. coluzzi* and *An. gambiae* along an east-west transect in Guinea Bissau in order to study the impacts of introgression in the population structure of these mosquito vectors, their ability to transmit malaria parasites and to respond to insecticide-based vector control measures. Microsatellite-based clustering analyses revealed a partitioning within *An. gambiae* into inland and coastal subpopulations, separated by a central region where *An. coluzzii* prevails. Whole genome sequencing revealed that coastal *An. gambiae* is a hybrid form characterised by an *An. gambiae* – like sex chromosome and massive introgression of *An. coluzzii* autosomal alleles. Insecticide resistance mutations were present in inland *An. gambiae* – east but absent (*ace-1* locus) or at very low frequency (*vgsc-kdr* locus) in the coastal hybrid form. Inland and coastal subpopulations also displayed differences in the human blood index and sporozoite rates. These results suggest that the genetic partitioning within *An. gambiae* may influence malaria epidemiology and control in this region.

Repellency and chemical composition of essential oils from plants used as natural mosquitoes insect repellent in communities in Oaxaca, Mexico

R. Perez-Pacheco and G. Estrada Perez
Instituto Politécnico Nacional, CIIDIR-IPN Oaxaca, Calle Hornos 1003, Col. Indeco Xoxocotlan, Oaxaca, 68000, Mexico; rafaelperezpacheco@yahoo.com

The mosquitoes are vectors of diseases like dengue, malaria, west nile virus, chicungunya and zika virus. The natural repellents constitute an alternative to reduce the risk of disease transmission and the excessive use of chemical insecticides. Take into account the empirical knowledge and use in communities in Oaxaca, Mexico, of plants as natural repellent against mosquitoes, was evaluated three concentrations of essential oils of *Aloysia tryphilla* Britton, *Bursera linaloe* Engelm, *Lantana camara* L, *Litsea glaucencens* Kunth, *Piper auritum* Kunth, *Porophyllum tagetoides* Kunth, *Tagetes lucida* Cav and *Satureja macrostema* Benth, using human bait technique recommended by the World Health Organization (WHO). Gas chromatography was used to identify major chemical compounds. The results indicate that *A. triphylla* oil in 50% concentration to cause the same effectiveness as registered with the synthetic repellent OFF 7.5% DEET, with a protection time of 150 minutes. The essential oil of *A. triphylla*, represents a potential alternative as a natural repellent. The oils of *L. camera*, *S. sacrostema*, *P. auritum* and *P. tagetoides* were statistically similar (α=0.05) with a time of protection 145.3, 135.9, 113.0 Y 101.5 minutes, respectively.

Foraging range of arthropods with veterinary interest: new insights for Afrotropical *Culicoides* biting midges (Diptera: Ceratopogonidae)

M.T. Bakhoum[1,2], M. Fall[1], M.T. Seck[1], L. Gardès[2], A.G. Fall[1], M. Diop[1], I. Mall[1], T. Balenghien[2], T. Baldet[2], G. Gimonneau[2], C. Garros[2] and J. Bouyer[2]
[1]Institut Sénégalais de Recherches Agricoles (ISRA), Laboratoire National d'Elevage et de Recherches Vétérinaire, Route du Front de Terre Dakar, Hann, BP 2057 Dakar, Hann, Senegal, [2]Cirad, UMR CMAEE, Montpellier cedex 5, France, 34398, France; thierno.bakhoum@cirad.fr

The identification of blood meal source of arthropod vector species contributes to the understanding of host-vector-pathogen interactions. The aim of the current work was to identify blood meal source in *Culicoides* biting midge species, biological vectors of internationally important arboviruses of livestock and equids, using a new ecological approach. We examined the correlation between blood meal source identified in engorged *Culicoides* females collected in a suction light trap and the available vertebrate hosts along four rings (200, 500, 1000 and 2,000 m) centered at the trap site and described the foraging range of the three main vector species of veterinary interest present in the study area, *Culicoides imicola*, *Culicoides kingi* and *Culicoides oxystoma*. The study was performed in four sites localized in the Niayes region of Senegal (West Africa) where recent outbreaks of African horse sickness occurred. Blood meal source identification was carried out by species-specific multiplex PCRs with genomic DNA extracted from the abdomen of engorged females collected during nine night collections for twenty-six collections. The four most abundant hosts present in the studied area (horse, cattle, goat and sheep) were surveyed in each ring zone. The blood meal source varied according to *Culicoides* species and host availability in each site. *C. oxystoma* and *C. imicola* females mainly fed on horses readily available at 200 m maximum from the trap location whereas females of *C. kingi* fed mainly on cattle, at variable distances from the traps (200 to 2,000 m). *C. oxystoma* may also feed on other vertebrates. We discuss the results in relation with the transmission of *Culicoides*-borne arboviruses and the species dispersion capacities.

Identification of blood meal sources in mosquito species (Diptera: Culicidae) using reverse line blotting in Aras Basin, Turkey

H. Bedir, B. Demirci and Z. Vatansever
Kafkas University, Department of Moleculer Biology and Genetics, 36100 Kars, Turkey; demirciberna80@gmail.com

In natural systems, detection of host preferences of mosquito species is import to clear up the ecological and evolutionary relationships between patogen-vector-host and control of vector-borne diseases. Despite the fact that the host preference in mosquitoes is genetical it also shows flexibility due to availability and accessibility of host species. In this project we aimed determination of host preference of the mosquito species of Aras Valley via blood meal analysis using the polimerase chain reaction (PCR) based reverse line blotting method. The proposed study aimed the determination of host preference of mosquitoe species in this location is done for the first time. DNA, extracted from blood fed mosquiotes collected from the field, were subjected to PCR with biotine labelled universal primers that amplify the 344 base pair cytochrome b gene region. Bloodmeal sources identified from 916 mosquitoes representing 7 different species by reverse line blotting (RLB). According to the reverse line blotting (RLB) hybridization results of 654 polymerase chain reaction positive individuals, it was observed that 323 (49.4%) of the mosquitoes fed on single hosts. From these single hosts, it was determined that 219 (68%) fed on human, 97 (30.1%) fed on cow, 5 (1.5%) fed on sheep, 1 (0.3%) fed on equus and 1 (0.3%) fed on pigeon. It was determined that other 331 (50.6%) individiuals prefered fed on multiple hosts and 290 (44.3%) of them took double, 30 (4.6%) of them took triple, 10 (1.6%) of them took quadruple and 1 (0.15%) them took quintette blood meal. In this study it was shown that blood feeding habits of mosquitos on human is quite high in this area and therefore, this may cause a high risk for transmission and spread of the mosquito-borne pathogens to humans. This may be the consequence of humans not taking any precaution to avoid or to reduce the contact with mosquitos. As a result, this study shows the urgent requirement of investigating mosquito-borne diseases in this area.

Insight on the larval habitat of Afrotropical *Culicoides* Latreille (Diptera: Ceratopogonidae) in the Niayes area of Senegal (West Africa)

M.T. Bakhoum[1,2], A.G. Fall[1], M. Fall[1], C.K. Bassene[1], T. Baldet[2], M.T. Seck[1], J. Bouyer[2,3], C. Garros[2] and G. Gimonneau[4]
[1]Institut Sénégalais de Recherches Agricoles (ISRA), Laboratoire National d'Elevage et de Recherches Vétérinaire, Route du Front de Terre Dakar, Hann, BP 2057 Dakar, Hann, Senegal, [2]Cirad, UMR CMAEE, Montpellier, France, 34398 Montpellier, France, [3]PATTEC coordination office, Addis Ababa, Ethiopia, P.O. Box 3243, Addis Ababa, Ethiopia, [4]Cirad, UMR INTERTRYP, Montpellier, France, 34398 Montpellier, France; thierno.bakhoum@cirad.fr

Certain biting midges species of the genus *Culicoides* (Diptera: Ceratopogonidae) are vectors of virus to livestock worldwide. *Culicoides* larval ecology has remained overlooked because of difficulties to identify breeding sites, methodological constraints to collect samples and lack of morphological tools to identify to species level of field-collected individuals. This study investigates larval habitats of *Culicoides* in horse farms in the Niayes area of Senegal using a combination of flotation and emergence methods. Among the three studied horse farms, three habitat types were found positive for *Culicoides* larvae: pond edge, lake edge and puddle edge. A total of 1,420 *Culicoides* individuals (519 ♂/901 ♀) belonging to 10 species emerged from the substrate samples. *Culicoides oxystoma* (40%), *C. similis* (25%) and *C. nivosus* (24%) were the most abundant species and emerged from the three habitat types while *C. kingi* (5%) was only retrieved from lake edges and one male emerged from puddle edge. *Culicoides imicola* (1.7%) was found in low number and retrieved only from pond and puddle edges. Larval habitats identified were not species-specific. All positive larval habitats were found outside the horse farms; no positive larval habitat inside the horse farms. This work provides original baseline information on larval habitats of *Culicoides* species in Senegal in an area endemic for AHSV, in particular for species of interest in animal health. These data will serve as a point of reference for future investigations on larval ecology studies and larval control measures.

Bioclimate envelope model for *Dermacentor marginatus* in Germany

M. Walter, K. Brugger and F. Rubel
University of Veterinary Medicine Vienna, Institute for Veterinary Public Health, Veterinaerplatz 1, 1210 Vienna, Austria; melanie.walter@vetmeduni.ac.at

Ticks are disease vectors widely distributed in Europe. Especially in human medicine they are important vectors for tick-borne encephalitis and Lyme borreliosis. In Germany, the most abundant vector is the hard tick *Ixodes ricinus*. The spatial distribution of other ticks is not yet known all over the country. Occurrence of *Dermancentor marginatus* is fragmented known, if at all. The tick was exclusively reported in the Rhine valley and adjacent areas. In contrast to other tick species it needs substantially warmer, drier and sunnier habitats. Particularly in focus of increasingly appearing zoonoses like rickettsioses, Q fever and tularemia, which can be transmitted by *D. marginatus*, a scientific evaluation of the potential distributional range is needed. On basis of geo-referenced tick locations and different environmental variables, like temperature and relative humidity, a statistical population model was implemented. The ecological niche for *D. marginatus* is calculated applying a dataset of 118 geo-referenced tick locations. It is described by 6 climate parameters based on temperature and relative humidity and another 6 environmental parameters including land cover classes and altitude. The final ecological niche is determined by the frequency distributions of these 12 parameters at the tick locations. This niche is used to estimate the habitat suitability and potential spatial distribution of *D. marginatus* in Germany by applying the BIOCLIM algorithm. The modelled potential distribution extends the already known distributional area along the Rhine river valley, to more northern and eastern parts of Germany.

Seasonal variation in abundance and biting behaviours of malaria vectors, *An. arabiensis* and *An. funestus* using climatic data in rural Tanzania

H.S. Ngowo[1,2], H.M. Ferguson[1] and F.O. Okumu[2,3]
[1]University of Glasgow, Institute of Biodiversity, Animal Health and Comparative Medicine, Graham Kerr Building, G12 8QQ, Glasgow, United Kingdom, [2]Ifakara Health Institute, Environmental Health and Ecological Science, Off Mlabani Passage, P.O. Box 53, Tanzania, [3]University of the Witwatersrand, School of Public Health, Johannesburg, 1 Jan Smuts Avenue, South Africa; hngowo@ihi.or.tz

Malaria transmission is highly influenced by climatic condition which drives the abundance and seasonal dynamics of *Anopheles* vectors. In particular, rainfall creates larval habitat for mosquito breeding while microclimatic conditions determine the survival and biting behaviours of *Anopheles* species. Data on mosquito abundance, rainfall, temperature and relative humidity from Southern-east Tanzania were analysed to investigates the contributions of microclimatic conditions in the changes of *Anopheles* biting behaviours after the widely use of Long Lasting Insecticide nets (LLINs). We conducted one year (616 night traps) longitudinal study in four villages in a malaria endemic area. Mosquito sampling were done using human landing catches (HLC) both indoor and outdoor location. Climatic information (temperature and relative humidity) were recorded simultaneous with mosquito collection. Rainfall data were calibrated for 14 days lagged and both current and lagged rainfall were used in the model to estimate maximum mosquito abundance. There is seasonal variation in abundance of *An. arabiensis* in the study area between indoors and outdoors (X_2=122.3, edf=6.59, P<0.001) and (X_2=104.7, edf=4.25, P<0.001), respectively. *Anopheles funestus* also has a seasonal variation over the year between indoors and outdoors (X_2=92.95, edf=6.08, P<0.001) and (X_2=75.68, edf=5.03, P<0.001), respectively. An increase in indoor temperature resulted to higher exophilic behaviours in *An. arabiensis* RR: 1.23 (95%CI [1.12-1.38], P<0.001) while *An. funestus* are less exophilic with an increase in indoor temperature RR: 0.91 (95%CI [0.67-0.1.24], P>0.05). These findings provide a combined estimate on how exposure to malaria vectors varies seasonally with variation in microclimatic conditions. Behavioural changes in *Anopheles* species is not only influenced by LLINs coverage but also changes in environmental condition around human dwellings.

Fauna of mosquitoes (Insecta: Diptera) of the Amur region, Russia

E.N. Bogdanova and V.P. Dremova
I.M. Sechenov First Moscow State Medical University, Novgorodskaja str. 22-1-16, 127572 Moscow, Russian Federation; nekton-zieger@mail.ru

Studies were performed in several climatic of Eastern Siberia, the Amur region. Blood-sucking mosquitoes in this region have a rich species compositions and very high numbers. In addition to the highly irritating when attack people, they are carriers of many vector-borne diseases, as well as cause allergic reactions and dermatitis. The species composition of mosquitoes in the Amur region, according to our data, accounted for 31 species: genera *Aedes* 27 species, *Culiseta* 2, *Culex* 1, *Anopheles* 1. Dominant in the different accounting items were *Ae. communis, Ae. pionips, Ae. punctor, Ae. cantans, Ae. excrucians, Ae. cinereus, Ae. riparius, Ae. vexans, An. hyrcanus*. In the zone of middle taiga are marked 18-21 species, on the southern taiga to 28 species, in the area subnemoral forest 8-13 species. When the distribution of mosquito species in their ecological characteristics in the middle taiga dominated oligothermophilic species, in the southern taiga and in subnemoral forest met oligo-, meso- and thermophilic species. Depending on the climatic conditions of the season, the proportion of mosquito species of different ecological groups were changed. Seasonal activity of adult mosquitoes began in III decade of May at the expense of overwintering females *Cs. bergrothi, Cs. alaskaensis* and *An. hyrcanus*. The flight of mosquitoes of genus *Aedes* started in the I decade of June and ended in the II decade of August-II decade of September. The highest number is observed in the II-III decades of June and I-II decades of August, depending on the item of observation. The maximum number of mosquitoes which attacking human arm in 20 minutes, reached 420 exemplars. The activity of mosquitoes had two peaks of high abundance: an evening at 18-22 hour and morning at 4-8 hour. The time of activity differed depending geographical points and climatic conditions.

Survival analysis of yellow fever mosquitoes (*Aedes aegypti*) at environmental fruit import conditions from South America to the Netherlands

A. Ibañez-Justicia, F. Jacobs, R. Van Den Biggelaar and C.J. Stroo
Centre for Monitoring of Vectors, Food and Consumer Product Safety Authority, P.O. Box 9102, 6700 HC Wageningen, the Netherlands; a.ibanezjusticia@nvwa.nl

Due the zika outbreak in South America, concern arises about the possibility of new introductions of the mosquito vector *Aedes aegypti* into the country with import of products. Since the foundation of the Centre for Monitoring of Vectors (CMV), exotic mosquitoes have been detected that were accidentally introduced at import locations of fruits. The Netherlands is the biggest importer from outside the European Union with almost 2.6 million tons of fruit and 367 thousand tons of vegetables, and fresh fruits arriving from South American harbors/airports. These products are maintained at different low temperatures and undergo trips of different lengths. To investigate the effect of low temperature on adult survival of *Aedes aegypti*, we set up an experiment at the quarantine facilities of the CMV. Mosquitoes were kept in environments similar to those used in fruit transport and submitted to temperatures comparable to those experienced during the process. ANOVA Fisher's protected least significant difference test was applied to the percentage of surviving mosquitoes at 4, 12, 14, 18 and 21 days and revealed low mean survival at lower temperatures (4, 8 and 12 degrees; $P<0.01$). Products that could allow import of yellow fever mosquitoes are products transported faster than 15 days (95% confidence interval upper bound) at constant temperatures of 12 °C or higher. This information can help to fine tuning the surveillance strategy at points of entry, focusing the efforts of invasive mosquito detection at the most likely associated locations and/or products.

Measurement of malaria vectors feeding behaviour: mosquito electrocuting trap as a tool for measuring mosquito host-choice and human biting rates

F.C. Meza[1], K.S. Kreppel[1,2], G.F. Killeen[1,3], H.M. Ferguson[2] and N.J. Govella[1]
[1]Ifakara Health Institute, P.O. Box 53, Ifakara, Tanzania, [2]University of Glasgow, Graham Kerr Building, G12 8QQ, United Kingdom, [3] Liverpool School of Tropical Medicine, Pembroke Place, L3 5QA, United Kingdom; fclement@ihi.or.tz

Host preference is commonly quantified based on examining mosquito blood meal source. This approach is however, biased toward the most abundant and accessible host species. Human landing catch, a method for quantifying number of mosquito bites that a human would be exposed to. Apart from the risk of exposing human subject to potentially infectious bites, it may also underestimating human biting rate. We evaluated for the first time tools which are exposure free, accommodating either a part or a whole body of host upon measuring both mosquito host-choice and human biting rate under full field condition. Mosquitoes were sampled using two different sizes of Mosquito Electrocuting Traps (MET): a small trap sufficient to encompass the lower legs of a seated person, and a large trap sufficient to encompass a calf or the entire body of a seated human. Comparisons of the number and diversity of mosquitoes collected were made between different trap sizes, different host baits and with the human landing catch. The experiment was conducted at outdoor catching stations in Kilombero Valley of Tanzania. A total of 27,107 female *Anopheline* mosquitoes were collected of which 16,124 (59.5%) were *An. arabiensis*, 3,571 (13.2%) *An. funestus* complex, 3,297 (12.2%) *An. coustani*, 1,847 (6.8%) *An. ziemanni* and the remaining 2,267 (8.4%) other *Anopheles*. Large METs baited with either cattle (RR=2.8) or humans (RR=2) collected at least twice as many *An. arabiensis* as the HLC. The small METSH also sampled slightly more *An. arabiensis* (RR=1.09) than HLC. Comparisons of MET collections made with human and cow baits indicated that *An. arabiensis* has no strong preference for either host (cow preference level = 0.44). Both sizes of METs have potential for assessing the malaria vector biting rates vectors. The MET demonstrated feasible and effective in the assessment of mosquito host-choice assay under competitive environment in full field setting which limits the biases introduced with the existing approach.

Wild rabbit burrows suitable hábitat for sandflies in the Ebro Valley (NE Spain)

J. Lucientes[1,2], P.M. Alarcón-Elbal[2,3], S. Delacour-Estrella[2], I. Ruiz-Arrondo[2,4], V. Oropeza-Velasquez[2] and R. Estrada-Peña[2]
[1]Instituto Agroalimentario de Aragón IA2, Calle Miguel Servet 177, 50013 Zaragoza, Spain, [2]Veterinary Faculty. University of Zaragoza, Animal Pathology, Calle Miguel Servet 177, 50013 Zaragoza, Spain, [3]Universidad Agroforestal Fernando Arturo de Meriño, Jarabacoa, Jarabacoa, Dominica, [4]Center of Rickettsioses and Arthropod-Borne Diseases, Hospital San Pedro-CIBIR, Logroño, Spain; jlucien@unizar.es

Recently it has been shown that the wild rabbit (*Oryctolagus cuniculus*) could be an appropriate reservoir of Leishmania infantum in Spain. The main vector of Leishmaniasis in Spain is *Phlebotomus perniciosus*. This species feeds on wild rabbit and can easily become infested. Indeed, studies of blood meal preferences show strong evidence that rabbits are contributing to the maintenance of a high sandfly population in the country. For this reason, it is interesting to study the phlebotomine sandfly fauna associated with this common lagomorph species. Periodic surveys were conducted over the past 15 years in order to study diseases of wild rabbit. These entomological surveys were carried out in three provinces (Navarra, Huesca and Zaragoza, located in the middle valley of the river Ebro, NE of Spain), covering a total of then localities. Exit-entrance traps were placed from dusk to down in the openings of the burrows. Besides several types of ectoparasites, especialy fleas, four species of sandflies were identified: *Phlebotomus* (*Phlebotomus*) *papatasi*, *Phlebotomus* (*Larroussius*) *ariasi*, *Phlebotomus* (*Larroussius*) *perniciosus* and *Phlebotomus* (*Larroussius*) *langeroni*. The most abundant and better distribuited species was *Phlebotomus perniciosus* in nine of the ten sampled localities. The second was *Phlebotomus papatasi* in five localities, and *Phlebotomus langeroni* in four, both species in the most arid área. By last *Phlebotomu ariasi* was captured also in theree localities, but only in Navarra province, in locations with higer humidity.

Distribution, abundance and ecology of ticks in Portugal mainland: data from five years of a surveillance program REVIVE

M.M. Santos-Silva, A.S. Santos, I. Lopes De Carvalho, R. De Sousa, T. Luz, P. Parreira, L. Chaínho, M.S. Gomes, N. Milhano, H. Osório, M.J. Alves, M.S. Núncio and REVIVE Workgroup
National Institute of Health Dr Ricardo Jorge, Centre for Vectors and Infectious Disease Research (INSA/CEVDI), Av Liberdade 5, 2965-575 Águas de Moura, Portugal; m.santos.silva@insa.min-saude.pt

REVIVE is a National Network for Vector Surveillance developed to enhance knowledge about vectors, their distribution and abundance, the impact of climate changes, their role as vectors and to detect invasive species with Public Health importance in a timely manner. REVIVE-Ticks started in 2011 and resulted from the networking of several public health institutions and is coordinated by INSA/CEVDI. Due to the first effects of climate change or to modifications in land usage, distribution of ticks is changing and its surveillance is considered of epidemiological relevance. This study reports a large scale survey set up from 2011 to 2015 for monitoring the activity of ticks, characterizing the species, its distribution, abundance, behavior, seasonal occurrence and activity peaks, and the host associations. During five years, 5,217 captures were performed in 180 counties representing 65% of mainland. More than 36,000 ticks were collected by flagging the vegetation and removed from vertebrate hosts, including man. 400 technicians were involved in the field work across the country. Up to date information were obtained on the activity of 13 autochthonous ticks namely *D. marginatus, D. reticulatus, H. punctata, H. lusitanicum, H. marginatum, I. canisuga, I. hexagonus, I. ricinus, I. ventalloi, R. annulatus, R. bursa, R. pusillus* and *R. sanguineus s.l.* One invasive tick species was recorded, an *Amblyomma* spp. found attached to a man. Distribution maps, climatic and ecological features are discussed for ticks that presented relative abundance upper than 1%. Since its implementation, REVIVE has contributed to increase the ecoepidemiological knowledge of vector species present in Portugal. This surveillance will call attention to any changes in abundance, diversity and ecological niches assisting in the identification of risk factors and the emission of public health alerts.

Culicoides species found near horses in Portugal which could be related to insect bite hypersensitivity (IBH)

P. Tilley[1*], *D.W. Ramilo*[1*], *S. Madeira*[1], *V.P. Pessoa*[1], *M. Branco Ferreira*[2] *and I. Pereira Da Fonseca*[1]

[1]*CIISA, Faculty of Veterinary Medicine, University of Lisbon, Animal Health, Av. Universidade Técnica, 1300-477 Lisboa, Portugal,* [2]*Faculdade de Medicina, Santa Maria Hospital, Universidade de Lisboa, Immunoalergology Unit, Internal Medicine Department, Av. Prof. Egas Moniz, 1649-028 Lisboa, Portugal; dwrramilo@hotmail.com;* [*]*These authors contributed equally to this work*

Insect bite hypersensitivity (IBH) is an allergic disease in horses caused by species of *Obsoletus* group, *C. nubeculosus* and *C. imicola*. The knowledge of *Culicoides* fauna present in each country and their ecological preferences can help in the establishment of control strategies for this and other diseases transmitted by this vector species. A study on IBH in Lusitano horses is being carried out, focusing on the *Culicoides* distribution near horse farms in Portugal. *Culicoides* were captured with 3 OVI traps in 13 horse farms in 9 regions of Portugal, including those where the IBH study is being carried out. *Culicoides* were identified to species by their wing pattern and when species identification was not possible, specimens were separated into different body parts followed by composed optical microscopy examination. Per night, an average of 50 specimens were collected in the trap sites localized above Tagus river and an average of 650 specimens were collected in the trap sites below Tagus river. A total of 29 different *Culicoides* species were captured near horse farms. Despite the high number of *Culicoides* species captured, the number of collected specimens depended on the location of the capture point in relation to Tagus river, being significantly higher in captures performed below that site. Although *C. nubeculosus* has not been found in the samples collected up to now, *C. puncticollis*, a very similar species, was found. As other species can also be involved in horse IBH, these preliminary results must be complemented with further captures and laboratorial studies on feeding preferences, since the presence of different *Culicoides* species in traps located in horse farms doesn't necessarily mean that they feed on horses. Funding: FCT-CIISA UID/CVT/00276/2013; VECTORNET OC/EFSA/AHAW/2013/02-FWC1; COST Action TD1303.

Faunistic inventory of mosquitoes (Diptera: Culicidae) and characterization of breeding sites in natural fountains of Barcelona city (Spain)

R. Bueno[1], A. Gomez[2], S. Navarro[3], J. López[4], L. Fernández[5], V. Peracho[6] and T. Montalvo[7]
[1]Laboratorios Lokimica, Departamento I+D, C/ Ferro, 3, 08038, Spain, [2]Agencia de Salud Pública de Barcelona, Servicio de Calidad e Intervención Ambiental, Av Príncep d'Asturies 63 2n.1a, 08012, Spain, [3]Agencia de Salud Pública de Barcelona, Servicio de Calidad e Intervención Ambiental, Av Príncep d'Asturies 63 2n.1a, 08012, Spain, [4]Laboratorios Lokimica, Departamento I+D, C/ Ferro, 3, 08038, Spain, [5]Agencia de Salud Pública de Barcelona, Servicio de Vigilancia y Control de Plagas Urbanas, Av Princep d'Astúries 63 3r.2a, 08012, Spain, [6]Agencia de Salud Pública de Barcelona, Servicio de Vigilancia y Control de Plagas Urbanas, Av Princep d'Asturies 63 3r.2a, 08012, Spain, [7]Agencia de Salud Pública de Barcelona, Servicio de Vigilancia y Control de Plagas Urbanas, Av Princep d'Asturies 63 3r.2a, 08012, Spain; rbueno@lokimica.es

Since the establishment of Asian tiger mosquito (*Aedes abopictus)* in Barcelona in 2005, a comprehensive management program has been implemented in the city with continuous tasks of monitoring and control the principal breeding sites in urban and public environments, like catch basins or small and artificial ponds. However, a high number of natural fountains are present in the municipality and most of them are placed in a particular interest environment with wild and forest conditions, called the Nature Reserve of Collserola. The aim of the study was to analyze the mosquito diversity in these 34 isolated water sources. Additionally, an exhaustive characterization of chemical and microbiological conditions of the these water collections has been done in order to deep into the knowledge of tolerance and preference factors of mosquitoes to select their breeding sites. A total of 318 larval exemplars of mosquitoes, belonging to the species *Culex pipiens*, *Culex laticinctus* and *Culiseta longiareolata*, has been collected and identified. It is important to remark that our surveys have allowed that *Cx. laticinctus* has been recorded for first time for Barcelona Province. Moreover, adult females of *Aedes albopictus* were also collected in some places. Analysis and discussions of likely breeding sites and flight range of Asian tiger mosquito in the forest area will be discussed through the use of tools related with Geographic Information Systems (GIS).

The seasonal dynamics of avian trypanosomes in mosquitoes

J. Radrova, O. Dolnik and M. Svobodova
Faculty of Science, Charles University in Prague, Parasitology, Vinicna 7, 12844 Prague, Czech Republic; radrova@natur.cuni.cz

To evaluate the role of *Culex* mosquitoes as potential vectors of avian trypanosomes, a long-term study is conducted in South Moravia, Milovice game preserve. Mosquitoes are captured by CO_2-baited CDC traps in 14-days intervals, identified, and kept in -20 °C until subsequent testing for *Trypanosoma* presence by Kinetoplastida-specific nested PCR assay. Between March and October 2014 and 2015, resp., 4,397 mosquitoes belonging to 12 species and five genera were caught. The prevailing species were: *Culex pipiens*, *Mansonia richiardii* and *Culiseta annulata*. *Mansonia* is peaking in the first half of July, while *Cs. annulata* two weeks later. *Cx. pipiens* has two peaks, in the first half of July and in the turn of July/August. After heavy rains in September 2014, *Aedes vexans* appeared in untypically high numbers. Mosquito females (740 pools) were tested by PCR; 96 pools were PCR positive. Total minimal infectious rate was 1.8% in 2014, and 4.5% in 2015, resp. From the seasonal point of view, the first *T. theileri*-infected *Culiseta annulata* was caught in end of May 2014, and *T. theileri*-infected *Anopheles plumbeus* in June 2015. Avian trypanosomes (*T. culicavium)* have been detected in *Cx. pipiens* in July. Resulting sequences represent three groups/species of *Trypanosoma: T. theileri s.l.*, *T. culicavium*, and *T. avium* s. l. Moreover, *Paratrypanosoma confusum*, *Crithidia* and *Leptomonas* spp. were detected. Mosquitoes are present on the locality from half of April till beginning of October, in comparison with previous years in relatively low numbers. *Trypanosoma* positive mosquitoes occurred only until late August. Transmission thus probably occurs only during a short late spring/summer period.

Biogeoclimatic factors affecting the spacial distribution and abundance of *Culicoides* (Diptera: Ceratopogonidae) in Castile-La Mancha, Spain

P.M. Alarcón-Elbal[1,2], R. Estrada[2], V.J. Carmona-Salido[2], S. Delacour-Estrella[2], I. Ruiz-Arrondo[2,3], C. Calvete[4,5] and J. Lucientes[2,4]
[1]Universidad Agroforestal Fernando Arturo de Meriño, Jarabacoa, Jarabacoa, Dominican Republic, [2]Facultad de Veterinaria. Universidad de Zaragoza, Patología Animal, Calle Miguel Servet 177, 50013 Zaragoza, Spain, [3]Center of Rickettsioses and Arthropod-Borne Diseases, Hospital San Pedro-CIBIR, Logroño, Spain, [4]Instituto Agroalimentario de Aragón IA2, Calle Miguel Servet 177, 50013 Zaragoza, Spain, [5]Centro de Investigación y Tecnología Agroalimentaria de Aragón (CITA), Montañana, Montañana, Spain; jlucien@unizar.es

Biotic (such as host distribution or movements) and abiotic factors (such as climate, land use or animal health systems) may affect distribution, abundance, seasonality and fecundity of *Culicoides* (Diptera: Ceratopogonidae). In this regard, statistical models are useful tools for identifying relationships between climatic and environmental factors, and the known distribution of vectors. This research, carried out with entomological data of the National Bluetongue Entomological Surveillance Program, was devoted to explore climatic, environmental and geographical factors affecting the spatial distribution and abundance of mammophilic *Culicoides* species in Castile-La Mancha, an autonomous community located in the middle of the Iberian peninsula. These species were *C. imicola, C. obsoletus/C. scoticus, C. circumscriptus, C. newsteadi, C. punctatus, C. pulicaris, C. nubeculosus* and *C. parroti*. In order to identify the distribution patterns of biting midges, abundance data was analyzed by PCA according to which *Culicoides* population is grouped into three components, which explain 89.9% of the total variation, mainly affected by vegetation cover and soil texture. The mean relative abundance of each species was determined, estimating the amount of spatial variation which can be explained by each of them. The percentage of variation explained ranged from 29.6% for *C. imicola*, distribution associated only with soil texture, to 88.4% and 98.5% for *C. parroti* and *C. nubeculosus*, respectively, both with its distribution associated with soil texture, use and vegetation cover, characteristics of livestock farms and certain climatic variables.

Twenty years of mosquito surveillance in the Czech Republic

F. Rettich[1], O. Šebesta[2] and I. Gelbič[3]
[1]Institute of Public Health, Vector control, Šrobárova 48, 100 42, Prague, Czech Republic, [2]Regional Public Health Authority of South Moravian Region, Sovadinova 450/12, 690 02 Břeclav, Czech Republic, [3]Biology Centre AS CR, Dept. of Entomology, Branišovská 31, České Budějovice, 370 05, Czech Republic; rettich@szu.cz

The two lowlands of the Czech Republic – the Morava/Dyje lowlands in Southern Moravia (Region 1) and the Labe lowlands in Central Bohemia (Region 2) – are areas, where mosquitoes (Diptera: Culicidae) occur annually, in some years in massive amounts. Vast breeding sites are located mainly on the banks of the Labe, Morava, and Dyje rivers, partly covered with remnants of floodplain forests. Important breeding sites of mosquito larvae are found in meadows of the inundation zones of the rivers mentioned above. The mosquito species distribution in the lowlands was also monitored in the Třeboň basin (a fish pond country) in Southern Bohemia (Region 3) and finally in the rolling Bohemian-Moravian Highlands in East Bohemia). Mosquito species distribution was monitored based on regular collections using four different methods: catches of mosquito larvae, catches of mosquito females on and around human volunteers, sweeping mosquito males around breeding sites, and EVS CO_2/light traps catches during the mosquito seasons for twenty years (year 2016 incl.). Altogether 35 mosquito species were identified in the long term study areas: 32 species in the Morava/Dyje lowlands 26 species in the Labe lowlands, 25 species in the Třeboň basin and 22 species in the Bohemian-Moravian Highlands. Species *Anopheles claviger*, *An. messeae*, *An. plumbeus*, *Aedes annulipes*, *Ae. cantans*, *Ae. cataphylla*, *Ae. cinereus*, *Ae. communis*,, *Ae. excrucians*, *Ae. geminus*, *Ae. geniculatus*, *Ae. leucomelas*, *Ae. rusticus*, *Ae. sticticus*, *Ae. vexans*, *Coquillettidia richiardii*, *Culiseta annulata*, *Cs. morsitans*, *Culex modestus*, *Cx. pipiens (*incl. *Cx. molestus* bioform), *Cx. territans* and *Cx. torrentium* were recorded in all regions (R1-R4) monitored. *Ae. flavescens*, *Ae. intrudens* and *Ae. rossicus* in the R1, R2, and R3, *Ae. caspius*, *Ae. dorsalis* in R1 and R2, *Cs. alaskaensis* R2 and R3 only, *An. hyrcanus*, *An. maculipennis s.s.*, *Cx. martini*, *Ur. unguiculata* in R1, *Ae. refikii* in R2, *Ae. pullatus* and *Cs. glaphyroptera* in R4 only.

Aedes cretinus Edwards in Turkey

F.M. Simsek[1], S.I. Yavasoglu[1], M. Akiner[2] and C. Ulger[1]
[1]Adnan Menderes University, Biology, Adnan Menderes Üniversitesi, 09010, Turkey, [2]Recep Tayyip Erdogan University, Biology, Rize, 53100, Turkey; fsimsek@adu.edu.tr

Aedes cretinus is a species of mosquito that has a restricted global distribution including Cyrus, Greece, Russia, Georgia and Turkey as well. Only few researches have been carried out about its biology and ecology due to the fact that it is seen in a restricted area worldwide and there is a not a dense population. No other data was obtained from this species that was recorded for the first time in Antalya-Gazipaşa around 1980s. Throughout this research, a study concerning larvae and adult sampling was done in Mediterranean, Aegean, and Black Sea regions from 2012 to 2015 with the aim of identifying different populations of *Ae. cretinus* in Turkey. These sampling was not only made from tree holes that were suitable for larval sampling but also from natural or artificial small breeding habitats. Adult specimens were caught by light traps and human bait traps. At the end of this present research, five more populations were identified in Adana, Mugla, Aydın, Artvin and Bursa in addition to Antalya, where the first record was obtained. It was observed that larvae samples were more intensely found to be in the holes of chestnut (*Castenea sativa)*, plane (*Planatus orientalis)* and cedar (*Cedrus libani)* trees compared to any other habitat. Furthermore, in Mugla population, it was observed that holes of plane and cedar trees were the most important breeding habitats for this species. While few females optained from light traps, a greater number of females caught in the human traps. It was also determined that adult females show high tendency of sucking human's blood. This study was funded by the Scientific and Technological Research Council of Turkey (project no. 1060841)

Distribution pattern and population dynamics of *Culex tritaeniorhynchus* Giles in Turkey

F.M. Simsek[1], S.I. Yavasoglu[1], M. Akiner[2] and C. Ulger[1]
[1]Adnan Menderes University, Department of Biology, Aydın, 09010, Turkey, [2]Recep Tayyip Erdogan University, Department of Biology, Rize, 53100, Turkey; fsimsek@adu.edu.tr

Culex tritaeniorhynchus Giles is a very important vector mosquito species that causes a variety of human diseases. It is the main vector of Japanese encephalitis in Southern Asia, and also has been detected in many regions of East and South Asia as infected with dengue, Rift Valley fever, sindbis, Getah and Tembusu viruses. It has been reported that it was able to be the vector of both *Brugia malayi* and *Wuchereria bancrofti* microfilariae. It is widely distributed throughout the Oriental region, extending into the Middle east, the Mediterranean and Afrotropical region. Although its records were given also from Turkey, there have been no studies determining the population distribution area of the species and demonstrating the seasonal population dynamics so far. In this study, the distribution areas of the species were revealed by conducting sampling researches in all zoogeographical regions of Turkey. In studies conducted between 2010 and 2015, different methods were used to sample the larvae and the adults of the species. The larvae samples were collected using standard larval dipper from the various breeding habitats. Adult samples were collected both in outdoor regions by light traps and animal traps and in indoor resting places like stables and houses by aspirator. As a result of the studies, it was found that distribution areas of the species in Turkey included South and West Anatolia and Thrace region. Adult samples were caught mostly with light traps in the sunrise and sunset time. Although number of individuals caught in animal traps varied according to the regions, it was more concentrated compared to the sampling made in indoor regions. In 2013 and 2014, seasonal population dynamism studies were performed in 5 populations selected from the distribution areas. In these studies, seasonal density fluctuations of the species were determined using light traps in March-October. According to the results obtained from light traps, it was determined that the highest population density of the species was reached in May and September within the year. This study was funded by The Scientific and Technological Research Council of Turkey (project no. 1060841)

First detection of *Wolbachia*-infected *Culicoides* in Europe: endosymbiont infection in the Palaearctic vectors of BTV and SBV

N. Pagès[1], F. Muñoz-Muñoz[2], M. Verdun[1], N. Pujol[1] and S. Talavera[1]
[1]IRTA, CReSA, Edifici CReSA, Campus de la UAB, 08193, Bellaterra, Spain, [2]Universitat Autonoma de Barcelona, Departament de Biologia Animal, de Biologia Vegetal i d'Ecologia, Campus de la UAB, 08193, Bellaterra, Spain; nonito.pages@irta.cat

Biting midges of the genus *Culicoides* (Diptera: Ceratopogonidae) biologically transmit pathogens that produce important diseases. No effective control technique is available to reduce neither *Culicoides* abundance nor their likelihood to transmit pathogens. Endosymbionts, particularly *Wolbachia*, represent powerful alternatives to control arthropods of health interest and reduce their disease transmission capacity. The presence of endosymbionts, *Wolbachia* and *Cardinium*, was tested in Spanish *Culicoides* populations. Regardless of geographical origin and landscape habitat endosymbionts were present in *Culicoides*, thus reporting the first detection of *Wolbachia*-infecting *Culicoides* in Europe. The putative Palaearctic vectors of bluetongue and Schmallenberg viruses were infected with *Wolbachia. Culicoides imicola* and *C. pulicaris* were infected with strains of A and B-Supergroups. *Culicoides* of the Obsoletus complex (that includes *C. obsoletus* and *C. scoticus)* were infected with strains of B-Supergroup. *Cardinium* strain of the C-group was detected in *Culicoides* Obsoletus complex. Both endosymbionts, *Wolbachia* and *Cardinium*, were detected in *Culicoides* species of minor epidemiological relevance as well. The presence of *Wolbachia* and *Cardinium* endosymbionts in *Culicoides* is expected to trigger new research on the control of *Culicoides*-transmitted diseases. Some *Wolbachia* strains are able to reduce vector fitness and vector's pathogen transmission in other arthropods. Therefore, *Wolbachia* would be a potential agent for population reduction and replacement strategies. The results of the present study could have an impact beyond *Culicoides* arena because successful *Wolbachia* transfection is possible even across genus and species barriers.

Distribution of some important mosquito species in Portugal within the framework of the national program for vector surveillance – REVIVE

H.C. Osório, L. Zé-Zé, M.J. Alves and REVIVE Workgroup
National Institute of Health Dr. Ricardo Jorge, Centre for Vectors and Infectious Diseases Research, Av. da Liberdade, 5, 2965-575 Águas de Moura, Portugal; hugo.osorio@insa.min-saude.pt

Knowledge of the diversity, actual distribution and ecological role of mosquito species is essential, to understand and evaluate the transmission risk of emerging pathogens, to develop effective strategies for disease prevention, as also to generate comprehensive information on the bionomics of species for successful development of any control strategy. In Portugal, a national mosquito surveillance program (REVIVE) is ongoing since 2008, which provide systematic collection of mosquitoes from several geographic regions and generate useful ecological data on species distribution, seasonal abundance and breeding habitats. In this study we present the geographic distribution of some medical important species recorded in the frame of REVIVE over the recent period 2011-2015, in which the mosquito fauna was systematically studied in 186 counties from 20 subregions (NUTS III). A total of 25 species were recorded in 3,617 and 6,043 adult and immature collections, respectively. In adult collections the most common species were *Aedes caspius*, followed by *Culex pipiens* and *Cx. theileri*. In immatures *Cx. pipiens*, *Culiseta longiareolata* and *Ae. aegypti* were the most abundant. The invasive vector species *Aedes aegypti* was only identified in the outermost Region of Madeira, where it is established since 2005. *Aedes albopictus* was never detected in Portugal. Surveillance and monitoring of the mosquito fauna allows timely detection of changes in abundance and species diversity providing valuable knowledge to health authorities, which may take control measures of vector populations reducing its impact on public health.

Analysis of spatial and temporal distribution of Ixodidae ticks in the Natural Reserve of Monte Pellegrino in Sicily, Italy

F. La Russa, M. Blanda, V. Blanda, R.M. Manzella, R. Disclafani, S. Scimeca, R. Scimeca, S. Caracappa and A. Torina
Istituto Zooprofilattico Sperimentale della Sicilia, Via Gino Marinuzzi 3, 90129 Palermo, Italy; alessandra.torina@izssicilia.it

Ticks (Acari: Ixodidae) are ectoparasites transmitting pathogens to humans, pets, wild animals and livestock. Tick activity is related to different environmental and climatic factors. Aim of this study was to carry out a spatial and temporal distribution analysis of free-living ticks in the Natural Reserve of Monte Pellegrino, in Palermo (Italy), a peri-urban area of the city attended by families, walkers, companion animals. A two-year monitoring (June 2012-May 2014) was carried out. Ticks were collected fortnightly by dragging method in six different sites (1. Sede Landolina, 2. Boschetto Airoldi, 3. Pineta Ex Scuderie Reali, 4. Sito Valdesi, 5. Castello Utveggio and 6. Gorgo S. Rosalia), with different environmental characteristics. Collected arthropods were identified basing on morphological keys. Tick population was analysed during all the months of the year and in the different collection sites. A total of 3,092 ticks (1,728 in the first year and 1,364 in the second one) was collected comprehending seven different species: *Ixodes ventalloi* (46.09%), *Hyalomma lusitanicum* (19.99%), *Rhipicephalus sanguineus* (17.34%), *R. pusillus* (16.11%) and, in less amount, *Haemaphisalis sulcata* (0.36%), *Dermacentor marginatus* (0.10%) and *R. turanicus* (0.03%). June 2012 (n. 324), April 2013 (n. 256) and January 2013 (n. 225) were the months with the highest numbers of ticks while the lower numbers of ticks were collected in August 2012 (n. 15), February 2013 (n. 34) and May 2014 (n. 38). The highest tick numbers were collected in the sites n. 2 and n. 5, the lowest in the site n. 4. In many cases, the same tick species was collected in sites showing similar environmental features. Tick density was related to altitude, land cover and vegetation. Monthly maps with circles proportional to the tick number were created using the geographical information systems. The study allows obtaining data on tick presence and distribution in the monitored area and constitutes a powerful surveillance tool. Authors thank Pippo Bono and Giulio Verro.

Morphological modifications on *C. imicola* palpus sensorial organs after exposure to pyrethroid molecules

D. Ramilo[1], R. Venail[2,3], G. Alexandre-Pires[1], T. Nunes[4] and I. Pereira Da Fonseca[1]
[1]CIISA, Faculty of Veterinary Medicine, University of Lisbon, Animal Health, Av. Universidade Técnica, 1300-477 Lisbon, Portugal, [2]Avia-GIS, Risschotlei 33, 2980 Zoersel, Belgium, [3]EID-Méditerranée, 165 avenue Paul-Rimbaud, 34184 Montpellier Cedex 4, France, [4]Faculty of Sciences, University of Lisbon, Microscopy Unit, Campo Grande, 1749-016, Lisboa, Portugal; dwrramilo@hotmail.com

Pyrethroid insecticides can be used to control *Culicoides* biting midges which are responsible for the transmission of several animal diseases. To perform their blood meal, *Culicoides* have specialized sensorial organs to detect the CO_2 exhaled by the vertebrate host. As pyrethroid molecules induce neurotoxicity, they may exert an effect on *Culicoides* sensorial organs of palpus, which detect CO_2. The objective was to evaluate morphological alterations in *C. imicola* sensorial organs of the 3rd palpus segment (sensilla basiconica) of alive and dead female after exposure to permethrin and deltamethrin. *C. imicola* females (n=47) were exposed for one hour to three different concentrations of permethrin (0.01%, 0.05% and 0.1%) and deltamethrin (0.0001%, 0.00025% and 0.001%) using the susceptibility test protocol of the World Health Organization adapted to this genus. Specimens were then divided into two groups (dead and alive after exposure) and a total of 101 individuals were prepared for scanning electron microscopy to observe the morphology of 3rd palpus segment sensorial organs, comparing with those of control group (n=54). All *C. imicola* analysed had morphological modifications on, at least, one of the 3rd palpus segment sensorial organs after permethrin and deltamethrin exposure at different concentrations. Pyrethroid insecticides induce physical modifications in *Culicoides* sensorial organs and, therefore, midges that survive after exposure to permethrin and deltamethrin may have difficulty in host localization, probably leading to their death by starvation. Further studies are ongoing. Funding: FCT Projects UID/CVT/00276/2013 and Pest-OE/AGR/UI0276/2014 and FCT fellowship SFRH/BD/77268/2011. Acknowledgments go to National Entomologic Surveillance Program of *Culicoides* biting midges (DGAV/FMV) and to its coordinator Prof. Fernando Boinas.

Risk of vector-borne diseases for the EU: entomological aspects

M. Braks[1], M. Goffredo[2], M. De Swart[1], G. Mancini[2], S. D'Hollander[3] and V. Versteirt[4]
[1]National Institute of Public Health and the Environment, Center Infectious Disease Control, A. van Leeuwenhoeklaan 9, 3720 BA Bilthoven, the Netherlands, [2]Istituto Zooprofilattico Sperimentale dell Abruzzo e del Molise, via Campo Boario, 64100 Teramo, Italy, [3]European Food Safety Authority, Via Carlo Magno 1A, 43126, Parma, Italy, [4]Avia-Gis, Risschotlei, 2980 Zoersel, Belgium; marieta.braks@rivm.nl

VectorNet was set-up as a joint initiative of the European Food Safety Authority (EFSA) and the European Centre for Disease Prevention and Control (ECDC) to strengthen both institutions' risk assessments activities on vector-borne diseases and to improve the preparedness of risk managers for these diseases in the European Union. VectorNet assists EFSA in its mandate requested by the European Commission to assess the risk of potential entry routes of vector-borne diseases into the EU. This presentation focuses on the entomological aspects of selected animal diseases and relevant zoonoses that may present a risk for Europe because of their possible introduction, re-introduction or further spread. A comprehensive worldwide list of vector species of the selected diseases was produced, including ticks, sandflies and *Culicoides* biting midges and mosquitoes. The list was compiled considering all species in which the pathogens were detected in field studies and/or in experimental infections, as reported in literature. To assess the risk of entry of pathogens via vectors, the general characteristics of the life cycles, infectious stages of these vector groups, their infection rates in nature and the different potential modes of transport of these infectious stages were explored. For enabling the assessment of the probability of introduction, spread and persistence in Europe, entomological data for the parameterisation of a vector-borne disease risk assessment model were provided.

A qualitative risk analysis of the distribution of African horse sickness in Nambia

D. Liebenberg, S.J. Piketh and H. Van Hamburg
North-West University (Potchefstroom campus), Unit for Environmental Sciences and Management, Private Bag X6001, 2520, Potchefstroom, South Africa; danica.liebenbergweyers@nwu.ac.za

African horse sickness (AHS) is one of the most lethal infectious, non-contagious, vector-borne disease of equids. The causative virus, African horse sickness virus is transmitted via *Culicoides* midges (Diptera: Ceratopogonidae). The disease has a seasonal occurrence that is influenced by environmental conditions that favour the breeding of *Culicoides* midges. During this study a risk analysis was performed to characterise the risk of the occurrence of AHS outbreaks in Namibia. Current methodologies for risk analysis focus on predicting the likelihood of distribution of diseases to non-endemic countries. However, to be able decrease the frequency of disease occurrences, it is important to understand the intra-country distribution. This is important for diseases such as AHS in countries where the disease is endemic – such as Namibia. For this risk analysis, principles of the World Organisation for Animal Health risk analysis were incorporated in combination with an ecological qualitative risk analysis, as well as a vector-borne livestock disease framework. A conceptual model was constructed which describes the relationships between identified sources, stressors, habitats and endpoints involved in the distribution of AHS. Stressors included: precipitation, temperature, humidity, normalised difference vegetation index, movement of horses and soil type. The immature stages of *Culicoides* development were estimated to contribute the highest to the occurrence on AHS outbreaks. A qualitative risk matrix was developed to estimate the risk of the occurrence of AHS outbreaks. The level of risk of a stressor was ranked according to analyses from previous studies. With the application of the risk matrix, the occurrence of AHS outbreak risk in a district can be estimated as low, intermediate or high. This is the first matrix to incorporate amongst other factors, anthropogenic activities and soil type as stressors to determine the risk of the occurrence of AHS outbreaks in Namibia. Although it is clear that limitations in the current risk matrix still exist, this investigation is of great importance for the development of an applicable tool for AHS management.

Towards a better understanding and performance improvement of species distribution models via a virtual vector

W. Tack, V. Versteirt, E. Ducheyne and G. Hendrickx
Avia-GIS, Risschotlei 33, 2980 Zoersel, Belgium; wtack@avia-gis.com

Species distribution models (SDMs) have become a vital tool in addressing environmental problems, such as managing rare and endangered species, forecasting the potential impacts of future environmental changes, controlling biological invasions and assessing vector-borne disease risk. Such SDMs use a variety of statistical methods to quantify species-environment relationships between occurrence data and a predictor set of environmental variables. The complex relationships are subsequently translated towards equations or a set of rules, and then used to spatially predict the most suitable/unsuitable areas. However, debate remains over the most robust modelling approaches for making projections as the performance of SDMs are influenced by data properties and species traits. In particular, sample size and prevalence are important factors influencing the performance of distribution models independently of range size. Small sample sizes weaken model performance and a biased sample prevalence (either low or high) negatively affects model predictions. An important question therefore is how sampling should be conducted to collect high-quality data and to obtain valid inferences about the species-environment relationships. This study examines the influence of sample size and prevalence on the performance of the models using simulated data. The advantage is the ability to test model performance against a known truth. A virtual species was constructed using the package 'virtualspecies' in R, entailing three major steps: (1) generating the probability of occurrence of a virtual species from a spatial set of environmental conditions; (2) converting the suitability into presence-absence with a probabilistic approach; and (3) extracting simulated observed data. We mapped the distribution of the virtual species in the European region using a spatial resolution of 30 arc-seconds (ca. 1 km/pixel). Various sampling schemes varying in sample size were used as training data set. The effect of balancing the training data set was studied by comparing models constructed using training data reflecting the natural sample versus balanced prevalence. Predictive models were VECMAP™ and the output was assessed using a set of accuracy measures.

Vector competence of *Culex pipiens* from Lebanon for West Nile fever virus and Rift Valley fever virus

R. Zakhia[1,2], N. Haddad[2], H. Bouharoun-Tayoun[2], L. Mousson[1] and A.B. Failloux[1]
[1]Pasteur Institute, Virology, Laboratory of Arboviruses and Insect Vectors, 25-28 Rue du Dr Roux, 75015 Paris, France, [2]Lebanese University, Laboratory of Immunology and Vector Borne Diseases, Faculty of Public Health, Fanar, El Metn, Lebanon; renee.zakhia@gmail.com

West Nile fever (WNF) is an arboviral infection caused by a flavivirus that is transmitted from infected birds to humans by a bite of infected mosquitoes. *Culex pipiens* is considered the most important vector for this disease. Although Lebanon is located in an endemic region, WNF has never been reported. However, a recent serological study, confirmed the exposure of Lebanese populations to West Nile virus (WNV) with a seroprevalence rate of 1.5% as confirmed by Plaque Reduction Neutralization Technique. Moreover, entomological monitoring showed the widespread of *Cx. pipiens* in Lebanon. These observations led us to assess the vector competence of local populations of *Cx. pipiens* towards WNV and Rift Valley fever virus (RVFV). *Cx. pipiens* eggs were collected during June 2015. 10 to 12 day age female mosquitoes were exposed to a blood meal containing WNV or RVFV at a titer of 1.6×10^8 PFU/ml and 1.33×10^7 PFU/ml respectively. Viral infection, dissemination and transmission were estimated at 3, 7, 14 and 19 days post-infection (dpi) using viral titration methods. Preliminary results showed that infection rate for WNV was very high at 3 dpi and remained high (90%) until 19 dpi. The dissemination rate reached 90% only at 14 dpi and the transmission rate exceeded 40% at 7 dpi then reached 90% at 14 dpi. In contrast, the infection rate for RVFV was around 40% at 3 dpi and reached 60% at 19 dpi. The dissemination rate was null at 3, 7 and 14 dpi then reached 20% at 19 dpi. Similarly, the transmission rate was null at 3, 7 and 14 dpi and did not exceed 10% at 19 dpi. These results show an obvious early transmission potential of WNV by Lebanese *Cx. pipiens* whereas transmission of RVFV occurred at a much lower rate and took more time to occur. This study highlighted the risk of WNV outbreak in Lebanon and the need to establish a national surveillance system to prevent the occurrence of diseases of major veterinary and public health concerns.

Geographical distribution of mosquito (Dipteria: Culicidae) in Morocco: first step to assess the transmission risk of arboviruses

A. Bennouna[1], G. Chlyeh[1], H. Elrhaffouli[2], F. Schaffner[3], W. Mahir[4], T. Balenghien[5] and O. Fassi Fihri[1]
[1]Hassan II Agronomy&Veterinary Institute, Madinat Al Irfane, Rabat, Morocco, [2]Faculty of Medicine and Pharmacy, Mohammed V Military Hospital, Rabat, Morocco, [3]Avia-GIS, Zoersel, Zoersel, Belgium, [4]Faculté de Médecine et de Pharmacie, Av. Med Belarbi El Alaoui, Rabat, Morocco, [5]UMR Cirad /Inra, Avenue Agropolis, 34398 Montpellier Cedex 5, France; amal397@yahoo.fr

Rift Valley fever is an acute viral zoonotic disease affecting both humans and domestic animals. The virus responsible for this disease belongs to the genus Phlebovirus, of the Bunyaviridae family. This work deals with an approach to assess the risk of the introduction, emergence and spread of arboviruses associated to mosquitoes. The assessment of the mosquitoes biodiversity in Morocco had led to do studies on almost all the territory of Morocco until today. Given the proximity of Morocco to Mauritania where the disease of Rift Valley fever has been confirmed, the aim is to make an inventory of current species in Morocco,to identify the most interesting sites and therefore, to achieve facilities of traps in the sites with the greatest potential. A crop of mosquito larvae has been carried out in several cities in Morocco, with an average of 140 different sites in total,during the whole period of the spring 2015 and 2016. The protocol of larval collection is respected and followed,in order to assess a crucial activity in the monitoring of the vectors of Rift Valley fever, especially *Culex pipiens*. The results obtained show a various diversity of species, depending on the preferences and ecological requirements for each type of breeding site. The identification has also revealed more than 4 species newly determined compared to previous work done in the Kingdom. Among those new species, a first record of the Asian tiger mosquito *Aedes albopictus* in Morocco was described based on both morphological identification and molecular analysis.A detection of pathogens using molecular and serological methods will either confirm or deny the presence of the virus by following validated protocols. All these data will be used to improve the monitoring programs and explore the microbiota of insects in relation with the vector competence and the emerging diseases.

Surveillance of mosquito-borne viruses with honey-baited FTA cards

N. Wipf[1,2,3], V. Guidi[2], O. Engler[4], E. Flacio[2], M. Tonolla[2], D. D. R. Guedes[5] and P. Müller[1,2,3]
[1]University of Basel, Petersplatz 1, 4003 Basel, Switzerland, [2]University of Applied Sciences and Arts of Southern Switzerland, Via Mirasole 22a, 6501 Bellinzona, Switzerland, [3]Swiss Tropical and Public Health Institute, Vector Control Group, Socinstrasse 57, 4002 Basel, Switzerland, [4]Federal Office for Civil Protection, Austrasse, 3700 Spiez, Switzerland, [5]CPqAM/Fiocruz-PE, Av. Moraes Rego, 50670-420 Recife, Brazil; nadja.wipf@unibas.ch

The threat of arboviral infections has recently reached continental Europe with several locally transmitted chikungunya and dengue cases. In southern Switzerland densities of the Asian tiger mosquito (*Aedes albopictus*), a competent vector of these viruses, have reached such levels that autochthonous transmissions have to be considered possible. Current surveillance efforts involve laborious processing of thousands of mosquitoes and are additionally hampered by the requirement of a constant cold-chain to preserve viral RNA. We evaluated a novel method that circumvents these limitations by exploiting the fact that infectious mosquitoes expectorate viruses in their saliva during sugar feeding. Within specialized traps, mosquitoes are allowed to feed on honey-baited, nucleic acid preserving FTA cards from which viral RNA can be directly purified and detected by RT-PCR. In preliminary laboratory experiments, viable yellow fever virus was found to be inactivated and preserved for more than one week on FTA cards. *Aedes aegypti* orally infected with dengue virus (DENV-2) were allowed to feed on honey-baited FTA cards. After 7 days, viral RNA was detected on 49% of the FTA cards that were exposed to infectious mosquitoes with DENV-2 positive salivary glands. In a field trail, honey-baited FTA cards were incorporated into different mosquito traps and set up in Recife, Brazil, where several arboviruses are endemic. Three different viruses – DENV-2, Western equine encephalitis, and chikungunya – were detected on weekly collected FTA cards. Based on the promising results, this novel technique is being implemented in the 2016 national arbovirus surveillance and the latest data will be shown at the e-SOVE meeting. The optimized method could serve as an early warning system and help to minimise the risk for mosquito-borne disease outbreaks in Switzerland and elsewhere.

VectorNet outcomes: mosquito vectors distribution maps and first round of field studies (2015)

F. Schaffner[1], D. Petric[2], V. Robert[3] and H. Kampen, & Vectornet Expert Contributors[4]
[1]FS Consultancy, Lörracherstr. 50, 4125 Riehen, Switzerland, [2]Faculty of Agriculture, Trg.D.Obradovica 8, 21000 Novi Sad, Serbia, [3]Institut de Recherche pour le Développement, 911 Av. Agropolis, 34394 Montpellier, France, [4]Friedrich-Loeffler-Institut, Südufer 10, 17493 Greifswald, Insel Riems, Germany; fschaffner.consult@gmail.com

VectorNet is a European Network for gathering and sharing data on the geographic distribution of arthropod vectors that can transmit human and animal disease agents. As a follow up of VBORNET, it is now funded jointly by the European Centre for Disease Prevention and Control (ECDC, Stockholm, Sweden) and the European Food and Safety Authority (EFSA, Parma, Italy). VectorNet maintains a vector distribution database for Europe, and provides this data through the publication of online quarterly-updated distribution maps of major vector species (among mosquitoes, ticks, sand flies and biting midges) in Europe. Besides published data and data sets directly shared by experts, the network also implements field studies aiming at filling gaps in vector's distribution knowledge. In 2015, a number of studies have been performed for mosquitoes, involving 18 partner institutions from 13 countries. In addition, opportune collaboration was established with WHO Regional Office for Europe, the international project MediLabSecure, and the regional project LOVCEN (Montenegro). Cross-sectional mosquito sampling was conducted at 41 areas and at more than 593 locations. The studies provided data to update the distribution maps for *Aedes aegypti*, *Ae. albopictus*, and *Ae. japonicus*. Additional significant result is the finding of *Culex tritaeniorhynchus* in Georgia. Further, no single mosquito species were found on Faroe Islands and in Iceland, despite the presence of suitable environment. This field programme has allowed to train more than 36 people from 18 different institutions and 13 countries. The training was provided during two MediLabSecure sessions in Serbia and Turkey, and during VectorNet field study start-ups, and focussed on the mosquito biology, mosquito sampling and identification techniques. Additionally, VectorNet is supporting the establishment of local multisector surveillance of invasive and native mosquito vector species and pathogens they transmit.

Implementation of entomological surveillance around imported cases of arbovirus diseases in the Balearic Islands (Spain)

M.A. Miranda[1], D. Borràs[1], R. García[2], M. Ruíz[1], C. Paredes-Esquivel[1], C. Barceló[1], M. Gumà[2] and M. Ramos[2]
[1]University of the Balearic Islands, Biology, Cra. Valldemossa km 7.5, 07122 Palma de Mallorca, Spain, [2]Conselleria de Salut, Direcció General de Salud Pública i Participació, c/ Jesús, 38ª, Palma de Mallorca 07010, Spain; ma.miranda@uib.es

The current events of arbovirus transmission such as dengue, chikungunya and recently zika viruses has led many countries to develop national programs on surveillance and response for arboviral diseases. In 2016, the Ministry of Health in Spain launched a national program that has been adopted by the different regions in Spain, including the Balearics Islands. The national program includes several levels of risk according to the presence/absence of potential vectors, such as *Aedes albopictus*, as well as the detection of arboviral imported cases or local transmission. The plan has been adapted to the situation and necessities in the Balearic Islands, where *Ae. albopictus* was detected in 2012 in Majorca, 2014 in Ibiza and 2015 in Minorca. An entomological surveillance protocol has been developed around arboviral disease imported cases taking into account three scenarios based on either the presence/absence of non-vector species and potential vector species for dengue, chikungunya and zika. The surveillance includes sampling of adults, larvae and eggs of mosquitoes in the residence and working place of the imported cases, as well as in public places around the case. A detailed spatial analysis of preferable habitats for *Ae. albopictus* (i.e. gardens, public parks, cemeteries) is conducted for each case by using GIS techniques and field sampling. The risk of local transmission of arbovirosis is estimated according to the epidemiological record of the case and the results of the entomological surveillance, taking into account the spatial analysis and other variables such as season, climatic conditions, as well as past and current mosquito control activities conducted at municipality level. According to the resulting level of risk, a list of recommendations is produced and provided to the responsible at municipal level. We present the results of the entomological surveillance and level of risk of local transmission of arboviral cases detected in 2016.

A nationwide surveillance on tick and tick-borne pathogens, 2011-2015, Portugal

R. De Sousa, A.S. Santos, M.M. Santos-Silva, T. Luz, P. Parreira, S. Bessa, M.S. Núncio, I. Lopes De Carvalho and REVIVE Workgroup
National Institute of Health Dr. Ricardo Jorge, CEVDI, Av. Liberdade no.5, 2965-575 Águas de Moura, Portugal; rsr.desousa@gmail.com

Active surveillance of ticks and tick-borne pathogens is crucial to assess the risks and understand the trends of tick-borne diseases. From 2011-2015, through the engagement of three institutions from the Portuguese Ministry of Health a nationwide tick and tick-borne pathogen surveillance program (REVIVE) was implemented on mainland Portugal. Here, we present the results of the five-year study relative to the molecular detection and identification of tick-borne pathogens such as *Rickettsia* and *Borrelia* in ticks collected from vegetation and different hosts across the country. Ticks collected in the studied period were identified and all the ticks collected in humans and 10% of ticks collected from animal hosts and vegetation were tested by PCR targeting *glt* A gene for *Rickettsia* spp and 5S (rrf)-23S (rrl) intergenic spacer for *Borrelia* spp. Of the total of 36018 collected ticks, 13 autochthonous tick species were identified from 180 Portuguese municipalities. The majority of ticks submitted were R. sanguineus (77.7%), followed by Hyalomma marginatum (4.8%). However the most common tick parasitizing humans was *I. ricinus*. From the total of 975 (2.7%) ticks from humans and 2976 (8.3%) ticks from animals or flagging from vegetation, screened by PCR, three *Borrelia* species (*B. lusitaniae, B. garinii, B. valaisiana, B. afzelii)* and eight Rickettsia species (*R. conori* i, *R. sibirica* mongolitimonae, *R. slovaca, R. massiliae, R. helvetica, R. monacensis, R. raoulti, R. aeschlimannii)* were detected. Co-infections of *B. lusitaniae*/*R. helvetica* and *B. lusitaniae*/*R. monacensis* were also found. *R. monacensis* and *B. lusitaniae* were the most prevalent *Rickettsia* and *Borrelia* species. Less pathogenic rickettsiae such as *R. monacensis, R. raoulti, R. massiliae* were more prevalent in ticks when compared with highly pathogenic rickettsiae such as *R. conorii.* Data showed that the distribution of *Rickettsia* spp. and *Borrelia* spp. found in ticks is homogeneous in mainland Portugal. This fact alert the physicians for the identification of different rickettsioses and *Borrelia* infections all over the country in different seasons of the year.

Monitoring of WNV, its vectors and reservoirs in Danube Delta Biosphere Reserve – Romania, one of the most important European gates for emerging diseases

M. Marinov[1], F.L. Prioteasa[2], E. Falcuta[2], V. Alexe[1], A. Dorosencu[1], J.B. Kiss[1], P. Tamiris[3], E. Veronesi[4] and C. Silaghi[4]
[1]Danube Delta National Institute for Research and Development, 165 Babadag street, 82011, Tulcea, Romania, [2]Cantacuzino Institute, Splaiul Independentei 103, 050096 Sector 5, Bucuresti, Romania, [3]3S.C. Aerocontrol Uav Srl, Str. Ilioara 15 A C, Bucuresti, Romania, [4]University of Zürich, Parasitology Dept., Winterthurestrasse 266a, 8057, Switzerland; mihai.marinov@ddni.ro

Since 2005 the Danube Delta National Institute (DDNI) together with Cantacuzino Institute, Bucharest have started to survey West Nile virus (WNV) in the Danube Delta Biosphere Reserve involving vectors and animal reservoir research. Data collected between 2005-2014 in Danube Delta shown the highest WNV seroprevalence in Europe among wild birds, several mosquito species involved in WNV transmission, their multiannual seasonal dynamics and infected bird ectoparasites. Moreover, overwintering surviving rates for *Culex pipiens* and WNV transmission in the absence of mosquitoes were investigated. In 2015, DDNI established a new department – Centre for the study of trans-border and emerging diseases and zoonoses – to build and accredit a centre that will include laboratories with different levels of biosecurity (the highest BSL will be 3.5 or even 4) which will allow biological samples from the Danube Delta to be handled and tested here by national and international teams. In the last years DDNI has initiated new surveys (e.g. funded by Romanian National Authority for Scientific Research) aiming to predict WNV infections using mosquitoes' natural infection and seroconversion rate in sentinel birds. Those parameters will be linked with the biodiversity and relative abundance of different vertebrate reservoirs in each investigated area using drones as well as with the microclimate particularities. In the institutional partnership AMSAR (Arbovirus Monitoring, SurveillAnce and Research – funded by the Swiss National Science Foundation; 2015-2017) have been shared the skills in bird and mosquitoes trapping, identification, sampling birds by different techniques, samples management and knowledge about ecology of vector and reservoir species for WNV with colleagues and students from veterinary universities in Serbia and Romania.

EU-ECDC/EFSA VectorNet project: distribution of sand fly species (Diptera: Psychodidae), community analysis and pathogen detection in Balkans

V. Dvorak[1], O.E. Kasap[2], G. Oguz[2], N. Ayhan[3], S. Vaselek[4], J. Omeragic[5], J. Pajovic[6], F. Martinkovic[7], O. Mikov[8], J. Stefanovska[9], D. Petric[4], D. Baymak[10], Y. Ozbel[11], J. Depaquit[12], V. Ivovic[13], P. Volf[1] and B. Alten[2,14]

[1]Charles University, Parasitology Department, Prague, Czech Republic, [2]Hacettepe University, Department of Biology, Beytepe, Ankara, Turkey, [3]Aix Marseille University, Medical Faculty, Virology Laboratories, Marseille, France, [4]Novi Sad University, Faculty of Agriculture, Novi Sad, Serbia, [5]University of Sarajevo, Veterinary Faculty, Sarajevo, Bosnia and Herzegovina, [6]Podgorica University, Veterinary Faculty, Podgorica, Montenegro, [7]University of Zagrep, Zagrep, Croatia, [8]National Center for Infectious and Parasitic Diseases, Sofia, Bulgaria, [9]Ss. Cyril and Methodius University, Faculty of Veterinary Medicine, Skopje, Macedonia, [10]National Institute of Public Health, Pristina, Zimbabwe, [11]Ege University, Faculty of Medicine, Department of Parasitology, Bornova, İzmir, Turkey [12]Universite de Reims, Faculty of Pharmacy, Reims, France, [13]Primorska University, Veterinary Faculty, Koper, Slovenia, [14]Hacettepe University, Institute of Science and Engineering, Beytepe, Ankara, Turkey; kaynas@hacettepe.edu.tr

In the framework of EU-ECDC/EFSA VectorNet project, a total of eight countries (Bosnia and Herzegovina, Montenegro, Croatia, Bulgaria, Macedonia, Serbia, Kosovo, Slovenia), 267 locations and 36 cities were studied by sand fly team with the aim of determined altitudinal and trans-sectional distribution of species, identification of species and detection of possible pathogens in Balkan. Sand flies were collected with total of 951 traps/night during the field missions. From this study, 12 species (*P. neglectus*, *P. tobbi*, *P. perfiliewi s.l.*, *P. papatasi*, *P. perniciosus*, *P. simici*, *P. alexandri*, *P. sergenti*, *P. balcanicus*, *P. mascittii*, *P. jacusieli*, *S. minuta)* were identified from a total of 9,096 specimens. The results shows that *Phlebotomus neglectus* (74%) major dominant species in Balkan countries and the species were collected from all countries with also *Phlebotomus tobbi* (10%). These two species comprise 84% of total abundance. Other species are: *P. perfiliewi s.l* (6.13%), *S. minuta* (3.56%), *P. perniciosus* (1.57%), *P. papatasi* (1.35%), *P. simici* (0.9%), *P. mascitti* (0.45%), *P. sergenti* (0.1%), *P. alexandri* (0.07%), *P. balcanicus* (0.03%) (found only in Montenegro). We determined altitudinal preferences of the species between 100-400 m with the optimum altitude of 200-300 m. We also calculated some of the important community parameters such as similarity, richness, species diversity, species evenness and dominancy for the countries. In pathogen detection studies, 2 (two) novel viruses in Bosnia and Herzegovina, and Macedonia, and *L. infantum* parasites in Bosnia and Herzegovina, Macedonia and Kosovo were determined.

Ecological niche and risk assessment of *Leishmania major* cutaneous leishmaniasis transmission in Algeria

R. Garni[1], M. Derghal[2], K. Belmokhtar[2], K. Benallal[1] and Z. Harrat[1]
[1]Pasteur Institute of Algeria, Parasitic Eco-epidemiology, Institut Pasteur d'Algérie, Route du petit Staoueli, Delly Brahim, Algiers, 16000, Algeria, [2]University of Sciences and Technologies Houari Boumedienne, Laboratoire d'Ecologie et Environnement, Université des Sciences et de la Technologie Houari Boumediene, BP 32, El Alia Bab Ezzouar, 16111, Algeria; ax19th@gmail.com

Despite the integrated vector control program launched in 2006, *Leishmania major* cutaneous leishmaniasis remains a major public health problem in Algeria with more than 6000 new cases recorded annually. The disease seems to spread toward the north of the country and could be attributed to the climate variability affecting the ecological niches of both the vector (*Phlebotomus papatasi)* and the reservoir hosts. Using geographic information systems in epidemiology is essential to understand relationships between the disease and the environment, and can help public health policy makers to fight the disease or adapt the control campaign stategy. The aim of this study is to predict the geographic distribution of *P. papatasi* using the maximum entropy approach (Maxent) and to assess the importance of environmental factors influencing the vector distribution. To determine the variables contributing to the dispersion of the vector, 84 environmental layers obtained from climate and GIS internet databases were used. For *P. papatasi*, 25 variables describing climate variability, altitude and vegetation were retained in the final model. The results show that moisture, rainfall and vegetation index contributed the most as predictor variables for *Phlebotomus papatasi*. The main parameters were the mean moisture index of coldest quarter (41%), the mean precipitation of November (7%) and mean Normalized Difference Vegetation Index (NDVI) of October (5%). In the other hand, the influence of the altitude in the the model is relatively low (3% of contribution). *P. papatasi* ecological niche shows highest suitability occurrence in the central and southeastern part of Algeria. Furthermore, by coupling this data with reservoirs host and human habitation maps, our study demonstrated that provinces of M'sila, Biskra and Batna in the east and Bechar in the west are at highest epidemic risk.

Harmvect – a simulation based tool for pathway risk maps of invasive arthropods in Belgium

F. Jansen[1], N. Berkvens[2], H. Casteels[2], J. Witters[2], V. Van Damme[2] and D. Berkvens[1]
[1]Institute of Tropical Medicine, Unit of Veterinary Biostatistics and Epidemiology, Nationalestraat 155, 2000 Antwerpen, Belgium, [2]Institute for Agricultural and Fisheries Research, Crop Protection Unit, Burgemeester Van Gansberghelaan 96, 9820 Merelbeke, Belgium; fjansen@itg.be

HARMVECT is a tool used for simulating the introduction risk of arthropods into Belgium. By simulating the arrival and entry of the arthropod into Belgium via the exploitable introduction pathways, the tool calculates pathway risk indices for arthropods threatening the Belgian food safety, animal and public health. These risk indices are then used to generate pathway risk maps. The tool is written in the open source R programming language and the Rstudio application 'Shiny' is used to build a user-friendly interface. The tool is designed in such a way that the user is guided in the process of breaking up the complex arrival of the assessed arthropod into its potential pathways, which in their turn are further subdivided into the smallest possible individual pathway-segments. These manageable and singular segments are used to specify which detailed data are needed to feed into the model. The size of the propagules being transported through the pathways is used to quantify the risks. The tool generates 3 pathway risk maps illustrating the spatial distribution and the magnitude of the arthropod's arrival. First, the point-of-entry (POE) pathway risk map is based on entry at the national borders and indicates concentrated hotspots. Second, the point-of-appearance (POA) pathway risk map is successive to the POE pathway risk map and points out the locations and level at which undetected propagules are exposed to the outside Belgian environment. The tool is also able to generate a total risk map combining the POA with a climate-suitability and host-availability map in order to provide information about the establishment capacity of the arthropod for Belgium.

Multiple mechanisms mediating insecticide resistance in *Aedes aegypti* from Madeira Island-Portugal

G. Seixas[1], L. Grigoraki[2], D. Weetman[3], J. Vicente[1], A.C. Silva[4], J. Pinto[1], J. Vontas[2] and C.A. Sousa[1]
[1]GHTM-Instituto de Higiene e Medicina Tropical-UNL, Rua da Junqueira 100, 1349-008 Lisboa, Portugal, [2]Molecular Entomology,IMBB, Foundation of Research and Technology, Hellas Nikolaou Plastira 100, GR-70013 Heraklion, Greece, [3]Department of Vector Biology, LSTM, Pembroke Place, L3 5QA Liverpool, United Kingdom, [4]Departamento de Saúde, Planeamento e Administração Geral, IASAUDE, IP-RAM, Rua das Pretas 1, 9004-515 Funchal, Portugal; gseixas@ihmt.unl.pt

Aedes aegypti (Linnaeus, 1762) is the main mosquito vector of dengue and zika. In 2005, *Ae. aegypti* was identified in Funchal, Madeira island. Despite the insecticide-based vector control, the species continued to expand throughout the island. The difficulties in controlling this mosquito population suggest presence of insecticide resistance (IR). Our study aims to characterize the phenotypic and genetic profile of IR in the *Ae. aegypti* population from Funchal. Susceptibility to 3 insecticide classes was evaluated with WHO bioassays. Metabolic resistance involvement was assessed with synergist and biochemical assays. Gene expression studies were made using microarrays, while knockdown resistance (*kdr*) mutations were genotyped in mosquitoes previously phenotyped as susceptible or resistant to pyrethroids. *Aedes aegypti* from Funchal was found resistant to all insecticides tested. Synergist assays with enzyme inhibitors significantly increased the mortality rates. Biochemical assays only showed the overactivity of esterases. Microarray analysis showed upregulation of genes associated with IR, mainly cuticle proteins (N=29) and cytochrome P450s oxidases (N=22). Of these, the most overexpressed Cyp9J32 and Cyp9J28 are known pyrethroid metabolizers. *Kdr* genotyping showed the presence of V1016I mutation with a frequency of 17% while F1534C is currently fixed. Insecticide resistance mediated by distinct mechanisms was identified in *Ae. aegypti* from Madeira. In addition to kdr and metabolic resistance, a third mechanism consisting of cuticle thickening may also be involved. Future control strategies should take into account the mechanisms underlying the resistance profile to discern which insecticides would be more effective against *Ae. aegypti* in Madeira. Funding: SFRH/BD/98873/2013; GHTM-UID/Multi/04413/2013.

Five years consecutive *Culicoides* spp. monitoring in Vienna to assess vector-free and transmission-free periods for bluetongue

K. Brugger, J. Köfer and F. Rubel
University of Veterinary Medicine Vienna, Institute for Veterinary Public Health, Veterinaerplatz 1, 1210 Vienna, Austria; katharina.brugger@vetmeduni.ac.at

Within the last few decades *Culicoides* spp. (Diptera: Ceratopogonidae) emerged Europe-wide as a major vector for epizootic viral diseases e.g. caused by bluetongue (BT) or Schmallenberg virus. In accordance with the EU regulation 1266/2007, veterinary authorities are requested to determine vector-free periods for loosing trade and movement restrictions of susceptible livestock. Additionally, the widely used basic reproduction number R_0 is optionally applied for transmission-free periods of vector-borne diseases. Values of $R_0<1$ indicate periods with no disease transmission risk. For the determination of vector-free period and R_0 a continuously operating daily *Culicoides* spp. monitoring in Vienna (Austria) was established. It covered the period 2009-2013 and depicts the seasonal vector abundance indoor and outdoor. The vector-free period lasted about 100 days inside stables, while less than five *Culicoides* were trapped outdoors on 150 days per season, i.e. winter half year. Additionally, the potential outbreak risk was assessed for BT and African horse sickness (AHS). For BT, a basic reproduction number of $R_0>1$ was found each year between June and August. The transmission-free periods, i.e. $R_0<1$, were notably higher (200 days). Contrary, values of $R_0<1$ were estimated for AHS during the whole period. Future BT and AHS outbreak risks were estimated by projecting R_0 to climate change scenarios. Therefore, temperature-dependent vector parameters were applied. While the mean summer peak values for BT increase from of $R_0=2.3$ to $R_0=3.4$ until 2100 (1.1/100 years), no risk for AHS was estimated even under climate warming assumptions. Restrictions to trade and movement are always associated with an economic impact during epidemic diseases. To minimize these impacts, risk assessments based on the vector-free period or transmission-free periods can essentially support veterinary authorities to improve protection and control measurements.

Spatial changes in West Nile virus circulation in Pianura Padana, Northern Italy, from 2013 to 2015

A. Albieri[1], R. Bellini[1], M. Calzolari[2], P. Bonilauri[2], M. Chiari[2], D. Lelli[2], M. Tamba[2], M. Dottori[2], G. Capelli[3], S. Ravagnan[3], P. Mulatti[3], C. Casalone[4], A. Pautasso[4] and C. Radaelli[4]
[1]Centro Agricoltura Ambiente 'G. Nicoli' / Sustenia S.r.l., Medical and Veterinary Entomology, Via Argini Nord, 3351, 40014 Crevalcore (BO), Italy, [2]IZSLER, Via Bianchi, 9, 25124 Brescia, Italy, [3]IZSVe, Viale dell'Università 10, 35020 Legnaro (PD), Italy, [4]IZSTO, Via Bologna, 148, 10154 Torino, Italy; aalbieri@caa.it

West Nile virus (WNV) is an emerging threat in Europe, with more than 1,000 human cases reported since 2010. After the first detection in Tuscany in 1998, WNV reappeared in north-eastern Italy in 2008-2009 affecting the regions of Emilia-Romagna, Veneto, and Lombardy, also reporting human cases. The virus was repeatedly detected in subsequent years, with different strains circulating over the years and in different areas of the involved regions. The environmental surveillance, particularly based on entomological and bird sampling, resulted highly efficient to early detect the circulation of WNV before the appearance of human and equine cases. From 2013 five neighboring regions of Northern Italy activated a similar entomological surveillance, allowing the monitoring of the whole Pianura Padana area (about 46,000 km^2). We compared WNV circulation maps from 2013 to 2015, using spatial interpolation and exploratory spatial data analysis methods (Kernel Density Estimation, Directional distribution, Mean center evaluation), to assess how the disease spread on the territory. WNV circulation area vastly expanded from 2013 to 2015: the KDE showed that the west front moved 82 km and 56 km from 2013 to 2014 and from 2014 to 2015, respectively, while mean center moved toward west of around 76 km from 2013 to 2015.

Tick-borne pathogens in ticks collected from domestic dogs and cats in Lithuania

J. Radzijevskaja, D. Mardosaite-Busaitiene, V. Sabunas and A. Paulauskas
Vytautas Magnus University, Vileikos 8, 44404 Kaunas, Lithuania; a.paulauskas@gmf.vdu.lt

Canine and feline vector-borne diseases are caused by a wide range of pathogens, which are transmitted by a variety of vectors, such as ticks, fleas, and mosquitoes. The spread of vectors and vector-borne pathogens in previously non-endemic areas due to climatic and ecological changes, and the increasing movement of pets can influence the epidemiological situation of vector-borne diseases in Europe. In Lithuania, *Ixodes ricinus* and *Dermacentor reticulatus* ticks are recognized as vectors of pathogens affecting humans, domestic and wild animals. The aim of this study was to investigate pathogens in ticks infesting domestic dogs and cats in Lithuania. A total of 257 adult ticks were collected from 46 dogs and 3 cats presented in three veterinary clinics during March-October 2015. *D. reticulatus* was the predominant species found on dogs (69%) and cats (81.6%), whereas *I. ricinus* were found in lower numbers (31% and 18.4%). Ticks were examined for the presence of *Anaplasma phagocytophilum* and *Babesia* spp. by using nested-PCR and sequence analysis of the msp4 and *18S rRNA* gene, respectively. *Babesia* parasites were detected in 10.5% of *D. reticulatus* and 7.6% of *I. ricinus* ticks from fifteen dogs, and in 28.6% of *I. ricinus* from two cats. Two *Babesia* species were identified in ticks infesting dogs: *Babesia canis* in *D. reticulatus* and *B. microti* in *I. ricinus*. *I. ricinus* from cats harbored *B. venatorum* and *B. canis*. *A. phagocytophilum* was detected in 9.2% of *I. ricinus* and in 27.1% of *D. reticulatus* ticks collected from fourteen dogs and in 3.9% of *I. ricinus* from one cats. Co-infections with both pathogens were detected in *D. reticulatus* ticks collected from 5 dogs. The high prevalence of pathogens found in ticks infesting pets demonstrate endemic occurrence of potentially zoonotic *Babesia* spp. and *A. phagocytophilum* and the risk of infection for dogs and cats in Lithuania.

The MediLabSecure network: capacity building in medical entomology in the Mediterranean and Black Sea regions

M. Picard[1,2], F. Gunay[1,2,3] and V. Robert[1,2]
[1]MediLabSecure, Medical and Veterinary Entomology, 911 Avenue Agropolis, 34394 Montpellier, France, [2]IRD, MIVEGEC/IRD 224-CNRS 5290, Montpellier University, France, 911 Avenue Agropolis, 34394 Montpellier, France, [3]Hacettepe University, Vector Ecology Research Group, Biology Department, Ecology Section, ESRL Laboratories, 06800, Beytepe-Ankara, Turkey; marie.picard@ird.fr

The EU-funded MediLabSecure project (2014-2017) aims at enhancing the preparedness and response to viral threats by establishing an integrated network of human virology, animal virology and entomology laboratories in 19 non-EU countries of the Mediterranean and Black Sea areas. The medical and veterinary entomology work package deals with awareness, integrated mosquito surveillance and risk assessment. After two-and-a-half years of existence, several actions were carried out. (1) Network of laboratories specializing in mosquitoes. A medical and veterinary entomology network of twenty laboratories is now established and active in the Mediterranean and Black Sea countries. (2) Mapping of the medical entomology expertise. An online tool has been set up to allow the dynamic visualisation of the laboratories as well as their field of expertise and the relevant facilities available. (3) Workshops on identification and surveillance of mosquito vectors species. Three five-day trainings were performed in Serbia (June 2015), in Turkey (September 2015) and in Tunisia (June 2016). A program of lectures, field work and lab activities was performed, allowing 50 participants from 19 countries to strengthen their knowledge in mosquito vectors of arboviruses and to improve regional networking. (4) 'One Health' networking activities. Intersectorial collaboration for arboviral disease surveillance in the One Health perspective is also encouraged through exchange of good practices and dissemination activities. During the two remaining years, the MediLabSecure WP medical and veterinary entomology will collaboratively carry on with these capacity building and networking activities. Above all, one of the challenges of this network is to harmonize the surveillance systems for mosquitoes in the Mediterranean and Black Sea Regions. MEDILABSECURE Project is supported by the European Union (DEVCO:IFS/21010/23/-194)

Bluetongue occurrences in Portugal and their relation with the entomological plan and official measures

S. Madeira[1], I. Pereira Da Fonseca[1], S. Quintans[2], R. Amador[2], D. Ramilo[1] and F. Boinas[1]
[1]CIISA, Faculty of Veterinary Medicine, University of Lisbon, Av. Universidade Técnica, 1300-477 Lisboa, Portugal, [2]Directorate-General of Food and Veterinary Medicine, Campo Grande, 50, 1700-093 Lisboa, Portugal; sara.pereira.madeira@gmail.com

Bluetongue virus (BTV) is transmitted by *Culicoides* causing an infectious, non-contagious disease of ruminants. It was first recorded in Portugal in 1956 and eradicated in 1960. After 44 years of epizootic silence, BTV re-emerged with BTV-4, which was considered eradicated in 2010, although sporadic occurrences only happened in 2013. In 2007 BTV-1 was detected in Alentejo, but the disease spread to northern regions and is still circulating in the country. The objective of this work is to describe bluetongue outbreaks in Portugal and their relation with the results of the entomological surveillance plan and the measures established by the Veterinary Official services. Measures implemented in Portugal are related to clinical, entomological, serological and virulogical surveillance, definition of seasonally restricted areas and implementation of a vaccination program and biosecurity measures. Databases including all the outbreaks, vaccinations, diagnostic lab results and entomological samples were available and were analysed for epidemiological inferences. This allowed the establishment of a relation between seasonality and distribution of BT occurrences and peak of collections of different species of *Culicoides*, and the establishment of a seasonally vector free period, that was adapted each year according to the evaluation of meteorological conditions. There was a positive evolution of the disease, related to the high vaccination coverage in previous years (animal coverage >70%), allowing the redefinition of the vaccination strategy implemented since 2012, which is now compulsory only in a risk area in the centre region of the country and voluntary in the remaining territory. Official measures are adopted according to the evolution of disease and risk evaluation based on surveillance results. The number of BT outbreaks decreased, and most of new cases occur in animals too young to be vaccinated or outside the limits of restriction zone, which shows that measures have been helping in disease control.

Molecular characterization of *Rickettsia* species in ticks from humans in Sicily (Italy)

V. Blanda[1], A. Torina[1], E. Giudice[2], R. D'Agostino[1], K. Randazzo[1], S. Scimeca[1], F. La Russa[1], R.M. Manzella[1], S. Caracappa[1] and A. Cascio[3]
[1]Istituto Zooprofilattico Sperimentale della Sicilia, Via G. Marinuzzi 3, 90129 Palermo, Italy, [2]Università di Messina, Polo Universitario Annunziata, 98168 Messina, Italy, [3]Università degli Studi di Palermo, 90127 Palemo, Via del Vespro, 129, 90127 Palermo, Italy; alessandra.torina@izssicilia.it

Rickettsiae (order Rickettsiales) are obligated intracellular bacteria transmitted by arthropod vectors. Spotted fever group of *Rickettsia* genus includes Rickettsiae agents of vector-borne rickettsioses. *Rickettsia conorii* was considered the main etiologic agent of Mediterranean spotted fever in the Mediterranean area. Molecular characterization of strains allowed identifying other *Rickettsia* species as agents of spotted fever in this area. The study aimed to *Rickettsia* species molecular characterization in ticks from humans in Sicily (Italy). The analysis concerned 42 ticks collected on humans from 2012 to 2013 in Messina district (Sicily, Italy) and identified by morphological keys. One of the patients showed rickettsiosis clinical manifestations. DNA was extracted from ticks and analysed by PCRs for *ompA*, *ompB* and *gltA* genes to detect *Rickettsia* spp. DNA. PCR products were sequenced. The following tick species were identified: *Rhipicephalus turanicus* (13 specimens), *Hyalomma lusitanicum* (11), *Rhipicephalus sanguineus* (7), *Dermacentor marginatus* (4), *Haemaphisalis punctata* (3), *Hyalomma marginatum* (2), *Ixodes ricinus* (1), *Rhipicephalus bursa* (1). Out of the 42 tick samples, 14 were positive to *Rickettsia* spp. Identified *Rickettsia* species included *R. conorii*, *Rickettsia aeschlimannii*, *Rickettsia massiliae* and *Rickettsia slovaca*. The first one was detected in *Rh. sanguineus* and *Rh. turanicus*. This latest tick was collected from the symptomatic patient. *R. aeschlimannii* was found in *H. marginatum*, *H. lusitanicum*, *D. marginatus* and *I. ricinus*. *R. massiliae* was detected in four *Rh. turanicus* ticks and in *Rh. sanguineus*, while *R. slovaca* was identified in *D. marginatus* and *Rh. sanguineus*. A great variety of zoonotic *Rickettsia* species was found in ticks collected from humans in Sicily, where transmission risk to humans is enhanced by the variety of tick species in the island. The authors thank Dr. Elda Marullo and Pippo Bono for technical support.

Emergence risk and surveillance of vector-borne diseases in Romania

G. Nicolescu[1], V. Purcarea-Ciulacu[1], A. Vladimirescu[1,2], E. Coipan[1], A. Petrisor[3], G. Dumitrescu[2], D. Saizu[4], E. Savin[4], I. Sandric[4], L. Zavate[4] and F. Mihai[4]
[1]'Cantacuzino' National Institute of Research, Medical Entomology, Spl. Independentei 103, Bucharest, 050096, Romania, [2]Army Center of Medical Research, Bucharest, 050094, Romania, [3]'Ion Mincu' University of Architecture and Urbanism, Bucharest, 050092, Romania, [4]ESRI Romania, Bucharest, 050098, Romania; gabrielamarianicolescu@yahoo.co.uk

Re-emergence of vector borne-diseases is a global process of major importance that appeared in the last years because of the climatic and other environmental changes generated especially by the human activities. The environmental changes influence by increasing of the distribution and abundance of vector populations and the contact of humans with vectors. Vector-borne diseases are important public health problems in Romania, and their re-emergence has to be prevented and control. The circulation of West Nile virus is endemic in Romania on large territories, and infections appear continuously as sporadic cases and occasionally severe outbreaks. There is a permanent risk of malaria re-appearance because of the simultaneous presence of the abundant local populations of anopheline vectors and the imported malaria cases. The Lyme borreliosis and tick-borne encephalitis have shown significant recent increase in incidence, mainly because of the changes in human behaviour in relation to the environment as in the rest of Europe. The surveillance of these diseases includes confirmation of human cases and their treatment when it is possible but this does not interrupt the disease transmission. The ecological surveillance is the essential activity for the prevention and removal of these diseases in Romania, taking into account that the main element in their appearance and epidemiology is the vector under the direct influence of environmental conditions. Our investigations put in evidence the ecological factors involved in spatial and temporal evolution of vector populations and the transmission of pathogens by them, and permitted the establishing of risk areas for every disease and the modalities of interruption of the transmission cycles by permanent monitoring and control of vectors applying integrated control programmes in anthropic ecosystems and habitats at risk.

Entomological surveillance and pathogen detection in mosquitoes in Greece (2010-2015)

S. Beleri[1], N. Tegos[1], G. Balatsos[1], D. Pervanidou[2], A. Vakali[2], A. Michaelakis[3] and E. Patsoula[1]
[1]National School of Public Health, Department of Parasitology, Entomology & Tropical Diseases, 196, Alexandras Avenue, 11521 Ambelokipi, Athens, Greece, [2]Hellenic Center for Disease Control and Prevention, Agrafon 3-5, 15123 Maroussi, Athens, Greece, [3]Benaki Phytopathological Institute, Department of Entomology and Agricultural Zoology, St. Delta 8, 14561 Kifisia, Greece; epatsoula@esdy.edu.gr

Entomological surveillance programs have been implemented in Greece, since 2010 aiming to record presence/absence of mosquito species, monitor populations and detect the circulation of arboviruses in mosquitoes at the local level. This data is useful to support decision making for targeted preventive measures. We present all the available data from the six years of the entomological surveillance. Using mainly CO_2 and Triple mosquito traps, collections were performed. Samples were examined and characterized to species level based on their morphological characters. They were then pooled by date of collection, location, species and examined for the presence of West Nile virus. *Culex pipiens* is the most widespread mosquito species and the main vector responsible for WNV transmission to humans. During the summers of 2014 and 2015, four and two imported dengue and for each year one chikungunya virus cases respectively were recorded. Focused entomological investigation was performed by setting ovitraps and BG-Sentinel traps, suitable for *Aedes* mosquitoes, so as to investigate the presence of potential vectors in the surrounding areas of the patients' residences, workplaces and hospitals of admission as well as the virus detection in mosquito pools. All pools containing *Aedes albopictus* were tested for the viruses and generated negative PCR results. *Ae. albopictus* is spreading in Greece and is a suitable vector for both viruses. Following recording of such cases, entomological surveillance is important to assess the risk of local transmission. It is essential to continue monitoring and perform annually vector surveillance studies, in stable stations, as they provide valuable entomological data and they can act as an early warning system for human cases and targeted public health interventions.

Immune gene evolution, ecological speciation and malaria transmission in African *Anopheles*

G.C. Lanzaro
University of California, Pathology, Microbiology & Immunology, 1089 Veterinary Medicine Drive, CA 95616, USA; gclanzaro@ucdavis.edu

We present a model under which the sister species *An. coluzzii* and *An. gambiae* have recently evolved via adaption to different larval habitats. The two species are principal vectors of malaria in Africa and occur together over most of their range. The two hybridize but hybrids appear to have reduced fitness and die out. We propose that hybrids have deficient immune systems and die due to infection with microbes present in the larval habitat. However, when impregnated bed-net usage increased the transfer of an insecticide resistance gene from *An. gambiae* into the genome of *An. coluzzii* resulted in an increase in the survival of hybrids. We discuss the presence of these hybrids on the biology of malaria transmission.

West Nile virus inhibits host-seeking response of the mosquito vector *Culex pipiens* biotype pipiens

C.B.F. Vogels[1], J.J. Fros[2], G.P. Pijlman[2], J.J.A. Van Loon[1], G. Gort[3] and C.J.M. Koenraadt[1]
[1]Wageningen University, Laboratory of Entomology, Droevendaalsesteeg 1, 6708 PB Wageningen, the Netherlands, [2]Wageningen University, Laboratory of Virology, Droevendaalsesteeg 1, 6708 PB Wageningen, the Netherlands, [3]Wageningen University, Biometris, Droevendaalsesteeg 1, 6708 PB Wageningen, the Netherlands; chantal.vogels@wur.nl

Host-seeking behaviour of mosquito vectors plays an essential role in transmission of pathogens, such as West Nile virus (WNV; family: Flaviviridae). WNV is transmitted among avian hosts by *Culex pipiens* mosquitoes, whereas mammals (including humans) are dead-end hosts. Manipulation of the mosquito's behaviour by WNV resulting in a stronger host preference towards avian hosts would, thus, be beneficial for WNV transmission. We hypothesized that WNV infection induces a stronger host-seeking response and a shift in host preference towards birds, in order to enhance WNV transmission by *Culex pipiens* mosquitoes. In this study, we first determined the effect of WNV infection on the fitness parameters flight activity, blood feeding propensity, and survival of *Culex pipiens* biotype *pipiens* females. Second, the host-seeking response of biotype *pipiens* females (control, mock-infected, and WNV-infected) for three odour stimuli (control, human, and chicken) was tested in a one-port olfactometer. Third, electrophysiological studies were done in order to compare the responses of antennal olfactory receptors of uninfected and WNV-infected biotype *pipiens* females for three host-derived odours (1% geranylacetone, 1% hexanoic acid, and 1% nonanal). In this presentation I will show that WNV inhibits the mosquito's host-seeking response, and does not induce a shift in mosquito host preference towards birds. Other fitness parameters were not affected by WNV infection. The reduced host-seeking response is likely not due to interference by WNV in the sensitivity of antennal olfactory neurons, but may be due to interference in the mosquito's central nervous system. Implications of these findings will be discussed in light of natural selection on WNV transmission dynamics.

Wolbachia pipientis in natural populations of mosquito vectors of dirofilaria immitis: first detection in *Culex theileri*

V. Mixão[1], A. Mendes[1], I. Maurício[1], M. Calado[1], M. Novo[1], S. Belo[1] and A.P.G. Almeida[1,2]
[1]Global Health and Tropical Medicine, Instituto de Higiene e Medicina Tropical, Universidade NOVA de Lisboa, R Junqueira 100, 1349-008 Lisboa, Portugal, [2]Center for Viral Zoonosis, Department of Medical Virology, University of Pretoria, Private Bag X20, Hatfield 0028, South Africa; palmeida@ihmt.unl.pt

Wolbachia pipientis (Rickettsiales: Rickettsiaceae) protects mosquitoes from infections with arboviruses and parasites. However, the effect of its co-infection on vector competence for *Dirofilaria immitis* (Spirurida: Onchocercidae) in the wild has not been investigated. Therefore, this study aimed to screen vectors of *D. immitis* for the presence of this endosymbiont, to characterize these, and to investigate a possible association between the occurrence of *W. pipientis* and that of the nematode. The presence of *W. pipientis* was assessed in the five mosquito potential vectors of *D. immitis* in Portugal, by polymerase chain reaction (PCR). The products of this reaction were sequenced, and all the matches with *W. pipientis* from nematodes were excluded, as some specimens were infected with *D. immitis*. For those samples carrying *w* Pip strain, haplotypes were determined by PCR-restricted fragment length polymorphism (RFLP). Results showed that *w* Pip was detected in 61.5% of *Culex pipiens* (Diptera: Culicidae) pools and 6.3% of *Culex theileri* pools. *w* Pip 16s rRNA sequences found in *Cx. theileri* exactly match those from *Cx. pipiens*, and this result was reinforced through a phylogenetic analysis, confirming a mosquito origin, rather than a nematode origin. Only *w* Pip haplotype I was found. No association was found between the presence of *w* Pip and *D. immitis* in mosquitoes and hence a role for this endosymbiont in influencing vectorial competence is yet to be identified. This study contributes to understanding of *w* Pip distribution in mosquito populations and, to the best of the authors' knowledge, is the first report of natural infections by *w* Pip in *Cx. theileri*.

On the trail of *Borrelia burgdorferi s.l.*: the key role of birds and lizards as reservoirs for the etiologic agent of Lyme borreliosis in Portugal

A.C. Norte[1,2], J.A. Ramos[2], L.P. Da Silva[2,3], P.M. Araújo[2], P.Q. Tenreiro[4], A. Alves Da Silva[3], J. Alves[3], L. Gern[5], M.S. Núncio[1] and I. Lopes De Carvalho[1]
[1]National Institute of Health Dr. Ricardo Jorge, Av. Padre Cruz, 1649-016, Portugal, [2]MARE-Marine and Environmental Sciences Centre, University of Coimbra, Largo Marquês de Pombal, 3004-517 Coimbra, Portugal, [3]University of Coimbra, Calçada Martim de Freitas, 3000-465, Portugal, [4]Instituto da Conservação da Natureza e das Florestas IP, Mata Nacional do Choupal, 3000-611 Coimbra, Portugal, [5]Institut de Biologie, Faculte des Sciences, rue emile-Argand 11, 2009 Neuchâtel, Switzerland; aclaudia.norte@gmail.com

Lyme borreliosis, caused by *Borrelia burgdorferi s.l.*, is the most prevalent vector-borne disease of moderate climates of the northern hemisfere. In Portugal, several *Borrelia* genospecies are present in questing ticks, but the most prevalent is *B. lusitaniae.* We studied the importance of birds, lizards and small mammals as reservoirs for *B. burgdorferi s.l.* in Portugal by collecting infesting ticks, and tissues (biopsies) and analysing them for *Borrelia* infection through nested PCR. A nationwide study revealed that *Ixodes ricinus* and *I. frontalis* were the most common ticks infesting birds, and that the most prevalent genospecies in ticks from birds was *B. turdi* (4.4%), a genospecies emerging in Europe. Other genospecies detected in ticks from birds included *B. valaisana* (3.7%), *B. garinii* (3.3%), *B. miyamotoi* (0.06%) and *B. bissettii* (0.06%). Thrushes were the most important birds in the enzootic cycle of *Borrelia*, due to high tick infestation rates, high tick *Borrelia* infection rates and also confirmation of reservoir competency though a xenodiagnostic experiment in *Turdus merula.* There was no evidence that birds acted as reservoirs for *B. lusitaniae*, but in a focal study in an enzootic area where this genospecies is highly prevalent, *B. lusitaniae* was detected in larval *I. ricinus* feeding on *Psammodromus algirus* (16.6%), on *Apodemus sylvaticus* (0.09%) and in one tail biopsy from *Podarcis hispanica* (5.9%), confirming the importance of mainly lizards, but also small mammals, as reservoirs for *B. lusitaniae.* This work was financially supported by the FCT (SFRH/BPD/62898/2009), by the University of Neuchâtel and by the National Institute of Health

Insecticide resistance in *Aedes aegypti* impacts its competence for chikungunya virus

L. Wang[1], P. Gaborit[1], M. Vignuzzi[2], R. Girod[1], D. Rousset[1] and I. Dusfour[1]
[1]Institut Pasteur of French Guinana, 23 Avenue Pasteur BP 6010, 97306 Cayenne Cedex, French Guiana, [2]Institut Pasteur of Paris, 28 rue du Dr Roux, 75724 Paris cedex, France; wanglanjiao@gmail.com

Aedes aegypti, is the principal vector of arboviruses such as dengue, chikungunya and zika. Insecticide resistance in *Ae. aegypti* seems to be associated with complex physiological changes, which may cause variations in vector competence including dissemination rate in mosquito body and transmission rate in saliva. In order to demonstrate a possible impact of insecticide resistance on virus multiplication, spread and selection in *Ae. aegypti*, four isofemale strains have been successfully isolated locally, they have similar genetic backgrounds except different resistance spectra to deltamethrin and different combinations of resistance mechanisms. They exhibit extreme phenotypes with resistance ratios of RR50>40 and <6 compared to the New Orleans reference susceptible strain. Resistance mechanisms are metabolic (Cytochrome P450) and/or target site mutations on the voltage-gated sodium channel (V1016I and F1534C). These isofemale mosquitoes and the susceptible reference mosquitoes have been infected orally with CHIKV, and then, saliva, midgut, head and salivary glands were collected at 3, 5, 7 and 10 days post infection (DPI). The virus detected by RT-qPCR, allowed us to evaluate both dissemination and transmission rates for each individual mosquito. Our data indicate that the isofemale mosquitoes used are as susceptible as the reference strain for oral infection with CHIKV, because the results of PCR for midguts samples were 100% positive. Furthermore, their dissemination rate (from midgut to head) is dependent on the mosquito strain used (highly resistant, moderately resistant and susceptible strains). In conclusion, insecticide resistance reduces dissemination rates of CHIKV (from midgut to head) in *Ae. aegypti* strains isolated in French Guiana. Further work will consist in comparisons of gene and protein expressions.

***Francisella*-like endosymbiont and *Ricketsia aeschlimannii* co-infection in a *Hyalomma marginatum* tick**

C.L. Carvalho, R. De Sousa, M.M. Santos-Silva, A.S. Santos, M.S. Núncio and I. Lopes De Carvalho
Instituto Nacional de Saúde Doutor Ricardo Jorge IP, Infectious Diseases, Av. Padre Cruz, 1649-016, Portugal; carina-l-carvalho@sapo.pt

Bacterial endosymbionts with significant homology to *F. tularensis*, a category A biowarfare pathogen, have been described in several tick species. *Francisella*-like endosymbionts (FLEs) have a worldwide distribution, are vertically transmitted by ticks. Recently they have been also detected in free-living small mammals. FLEs pathogenicity to humans is uncertain and their effect, if any, on vector competency and in the transmission cycle of *F. tularensis* or other tick-borne agents is still unknown. A total of 341 ticks from several species, including *Dermacentor marginatus* (n=10), *Dermacentor reticulatus* (n=19), *Hyalomma marginatum* (n=8), *Hyalomma lusitanicum* (n=8), *Ixodes ricinus* (n=266), *Rhipicephalus pusillus* (n=1) and *Rhipicephalus sanguineus* (n=29), were collected from the vegetation and different mammalian hosts, including humans, dogs and deers. Ticks were assembled by a National surveillance program on arthropod vectors (REVIVE) and other ongoing projects. Ticks were tested by conventional PCR targeting *tul4* gene for *F. tularensis* and citrate synthase for *Rickettsia* spp. followed by sequencing analysis. All ticks tested negative for *F. tularensis* but nucleotide sequences with high nucleotide similarity to FLEs were identified in 2.3% of the questing ticks, including *D. reticulatus* collected from humans (0.58%) and dogs (1.17%), and *H. marginatum* (0.29%) and *H. lusitanicum* (0.29%) from humans. Furthermore, *Rickettsia aeschlimannii*, a spotted fever group pathogen, was detected in one of the *H. marginatum* specimen infected with FLEs (0.29%). This particular co-infection had never been reported before in *H. marginatum* ticks. This study highlights the need to clearly understand the role of FLEs, if any, in the transmission of tick-borne agents. This work was partially supported by an FCT grant: SFRH/BD/79225/2011 and REVIVE program.

Implicating vectors of Schmallenberg virus in Europe

N. Pagès[1], M. Goffredo[2], S. Carpenter[3], S. Talavera[1], F. Monaco[2], J. Barber[3], V. Federici[2], M. Verdun[1], N. Pujol[1], A. Bensaid[1] and J. Pujols[1]
[1]IRTA-CReSA, Centre de Recerca en Sanitat Animal, Campus de la UAB, 08193, Bellaterra, Spain, [2]Istituto Zoprofilattico Sperimentale dell'Abruzzo e del Molise 'G. Caporale', Campo Boario, 64100, Teramo, Italy, [3]The Pirbright Institute, Ash Road, GU24 0NF, Pirbright, United Kingdom; nitu.pages@gmail.com

In autumn 2011, Schmallenberg virus (SBV) was described in cattle from Germany. The virus produces a disease characterized by a rather mild or subclinical infection with short viraemia, followed by abortion and congenital deformities when pregnant ruminants are infected. Probably, asymptomatic infections lead to a silent epizootic that rapidly spread across Europe. SBV is a novel *Orthobunyavirus*, family *Bunyaviridae*, placed within the Simbu serogroup. Other closely related viruses of the Simbu serogroup have been isolated both from mosquitoes and *Culicoides*. *Culicoides* biting midges were proposed as SBV vectors based on the presence of viral RNA in field collected *Culicoides*, mainly species of the Obsoletus complex. Experimental infections with mosquitoes and *Culicoides* were performed to assess their susceptibility to SBV infection. When orally infected with SBV, mosquito laboratory colonies were refractory to infection. However laboratory colonies of *C. nubeculosus* and *C. sonorensis* were susceptible with low infection rates (IR). Thereafter, *Culicoides* field populations were tested for SBV oral infection in Italy, Spain and the United Kingdom. Field collected *C. imicola* and Obsoletus complex (this including *C. obsoletus* and *C. scoticus)* evidenced high susceptibility for SBV infection. High disseminated infection rates and isolation of infectious virus from head homogenates suggest that those species may have high transmission ability and could act as efficient SBV vectors.

Preliminary results on the microbiota of *Ixodes ricinus* ticks of North-East of Italy

S. Ravagnan, E. Mastrorilli, S. Cazzin, A. Salviato, A. Milani, F. Montarsi, C. Losasso, A. Ricci, I. Monne and G. Capelli
Istituto Zooprofilattico Sperimentale delle Venezie, Parasitology department, Viale dell'Università 10, 35020 Legnaro, Padua, Italy; gcapelli@izsvenezie.it

Ixodes ricinus is the most prevalent tick in Europe and it is the vector of a broad range of bacterial pathogens affecting animals and humans. *Ixodes ricinus* can also hold bacteria complex microbial community (microbiota) that can develop both mutualistic and pathogenic relationships with the host. This study aimed to characterize microbiota in single adult ticks using targeted amplicon sequencing (16S rDNA). Ten *Ixodes ricinus* adult ticks were collected from vegetation by dragging in Belluno province in 2015, an area in the North-East of Italy where tick-borne pathogens are known to be endemic. The V1-V2 and V3-V4 hypervariable regions of *16S rDNA* gene were amplified for each tick and sequenced by Illumina MiSeq platform. Overall, 308 genera belonging to 20 bacteria phyla were found. Proteobacteria was the dominant phylum, followed by Actinobacteria, Firmicutes and Bacteroidetes. Among the Anaplasmataceae family, medically important arthropod-borne pathogens such as *Anaplasma* and *Ehrlichia* genera, were identified. Species belonging to these genera are considered pathogenic for both animals and humans. Other genera, which may hold pathogenic or opportunistic species, such as *Mycobacterium*, *Clostridium*, *Burkholderia*, *Rhodococcus* and *Bacillus*, were identified. Interestingly, *Wolbachia* genus was identified as well. The endosymbiotic *Wolbachia* can cause a variety of reproductive alterations in their arthropod hosts and can affect the transmission dynamics of pathogenic bacterial species. Among Gammaproteobacteria class, *Rickettsiella* genus is able to impact stage development and survival of their natural arthropod hosts. The targeted 16S rDNA regions used allowed to unravel a high bacterial diversity including rare taxa. The simultaneous detection of pathogenic and symbiotic taxa can have important applications in diagnosis and ecological studies. How the microbial community dynamic of *Ixodes ricinus*, as well as other arthropods, can modulate the vector susceptibility to pathogens and their transmission is a future scientific challenge.

Recycling of parasitic nematodes *Romanomermis culicivorax*, in breeding sites of mosquito larvae

R. Perez-Pacheco[1], N. Bautista-Martínez[2], G. Flores Ambrosio[1] and S. Martínez Tomas[1]
[1]Instituto Politécnico Nacional, CIIDIR-IPN Oaxaca, Calle Hornos 1003, Col. Indeco Xoxocotlan, Oaxaca, 68000, Mexico, [2]Colegio de Posgraduados, Instituto de fitosanidad Campus Montecillo, Montecillo, Estado de Mexico, 56230, Mexico; rafaelperezpacheco@yahoo.com

The set up adult nematodes in natural breeding of mosquito larvae is an alternative to guarantee long time of mosquito population control. We used four breeding sites of mosquito larvae *Cx. quinquefasciatus* with substrate of 10 cm of a mixture of (gravel and sand) and two breeding sites with sand as substrate of 10 cm. In two breeding sites with (sand and gravel) 200 adult nematodes (100 males and 100 females) were set up and the other four breeding sites two with (gravel and sand), and two with (sand) 100 nematodes (50 males and 50 females) were set up. 19 days after of set up nematodes, were added 500 mosquito larvae per breeding sites and another 20 larvae in a sentinel trap, after five days, the trap were removed, and the larvae were dissected to determine levels of parasitism and the means of infestation (number of nematodes per larva). To determine infectivity of nematodes (percentages parasitism, PP) and the means of infestation (MI, number of nematodes per larvae), in the first 100 days every week were added, after of 100 day were added every two week and 250 days after were added every three weeks. The study continued during 468 days after set up nematodes. Were recorded percentage of parasitism of 30 to 45%. In the breeding sites with (gravel and sand), the highest levels of infestation were presented. The level of parasitism remained high in the breeding sites per 468 days

The role of biting midges in avian trypanosomes transmission

M. Svobodová[1], O. Dolnik[1], I. Čepička[2] and J. Rádrová[1]
[1]Charles University in Prague, Faculty of Science, Department of Parasitology, Viničná 7, 12844 Prague 2, Czech Republic, [2]Charles University in Prague, Faculty of Science, Department of Zoology, Viničná 7, 12844 Prague 2, Czech Republic; milena@natur.cuni.cz

Avian trypanosomes have been shown to form three major clades: *T.* cf. *avium*, *T. corvi*/*T. culicavium*, and *T. bennetti*. Confirmed vectors for the first two clades are blackflies (Simuliidae), hippoboscids (Hippoboscidae) and mosquitoes (Culicidae); vectors of *T. bennetti* remain unknown. Our aim was to clarify the role of biting midges as vectors of avian trypanosomes. Based on the results of our extensive survey of ornithophilic bloodsucking insects in the wild, we supposed that midges do not act as vectors of avian trypanosomes since no infected midge has been found. Laboratory-bred *Culicoides* nubeculosus have been artificially fed on different trypanosome strains, showing high susceptibility for *T. bennetti* (90-100% infected, heavy infections in 50-87% of individuals), and for *T. avium* as well (85% infected, heavy infections in 55%). On the other hand, *C. nubeculosus* was not permissive for *T. culicavium*. Infectivity of trypanosome stages from midges has been tested using canaries (*Serinus canaria*). *T. bennetti* strains were infective only after subcutaneous inoculation while *T. avium* transconjunctivally or perorally. Parasites were localized in the abdominal midgut and hindgut, resp., of the insect vector. Previous studies on midges transmission potential used trypanosomes that were not molecularly characterized. We confirm the permissiveness of *C. nubeculosus* biting midges for avian trypanosomes from both *T. bennetti* and *T. avium* clades; natural vector species of *T. bennetti* remain to be confirmed.

Co-circulation of phleboviruses and *Leishmania infantum* in a zoonotic visceral Leishmaniasis focus located in Central Tunisia

W. Fares, K. Dachraoui, W. Barhoumi, S. Cherni, W. Fraihi, S. Sakhria, M. Derbali, Z. Zoghlami, T. Ben Slimane, I. Chelbi and E. Zhioua
Institut Pasteur de Tunis, Laboratory of Vector Ecology, 13 Place Pasteur, 1002 Tunis, Tunisia; elyes.zhioua@gmail.com

In the Western Mediterranean basin, sand flies are vectors of several pathogens affecting humans and animals. In this study, we aimed to investigate phlebovirus and *Leishmania* parasites circulating in a focus of zoonotic visceral Leishmaniasis (ZVL) located in arid bio-geographical areas of Central Tunisia. Sand flies were collected by using CDC light traps in the villages of Felta and Saddaguia located in Central Tunisia during September-October 2014 and 2015. Live sand flies were pooled with up to 30 specimens per pool and screened for phlebovirus and *Leishmania* by nested PCR in the polymerase gene and kinetoplast minicircle DNA, respectively. Dead sand flies were identified morphologically to species level. A total of 3,191 and 2,229 sand flies were collected in 2014 and 2015, respectively. Among 117 pools of sand flies collected in 2014, 4 were positive for phleboviruses yielding a minimum infection rate (MIR) of 0.12% (4/3191). Indentified phleboviruses were: 1 Utique virus (UTIV), 1 sandfly fever Sicilian virus (SFSV), 1 Saddaguia virus (SADV) and 1 Toscana virus (TOSV). While UTIV and SFSV were detected in pools of males, TOSV and UTIV were detected in pools of female sand flies. Of a total of 52 pools of female sand flies, 3 were positive for *Leishmania infantum* DNA on the basis of PCR amplification size, yielding a MIR of 0.1% (3/1370). Among 80 pools of sand flies collected in 2015, 3 were positive for phleboviruses consisting of 2 pools of TOSV and 1 pool of SFSV yielding a MIR of 0.13% (3/2,229). All phleboviruses were detected in pools of female sand flies. Among 36 pools of female sand flies, 1 was positive for *L. infantum*, yielding a MIR of 0.1% (1/995). No co-infected pools with phlebovirus and *L. infantum* were detected. Sand flies of the subgenus *Larroussius* were predominant in this ZVL focus. Our results provide good evidence of the co-circulation of phleboviruses and *L. infantum* in this ZVL focus.

DNA barcoding of farm-associated mosquitoes (Diptera: Culicidae) in southern England: species identification and discovery of cryptic diversity

L.M. Hernández-Triana[1], V.A. Brugman[2], N.I. Nikolova[3], L. Thorne[1], M. Fernández De Marcos[1], A.R. Fooks[1] and N. Johnson[1]
[1]Animal and Plant Health Agency, Virology, Woodham Lane, New Haw, Addlestone, Surrey, KT15 3NB, United Kingdom, [2]London School and Tropical Medicine Inst., Keppel Street London, WC1E 7HT, United Kingdom,-[3]Biodiversity Institute of Ontario, 50 Stone Road East, Guelph, ON, Canada, N1G 2W1, Canada; luis.hernandez-triana@ahvla.gsi.gov.uk

A number of UK mosquito species (e.g. *Culex pipiens* and *Aedes vexans)* are vectors of pathogens of medical and/or veterinary importance. Thus, correct species identification is critical for disease control programs. However, morphology-based taxonomy of mosquitoes is difficult for some species complexes. In this paper, we evaluate three DNA extraction methods to obtain rapid COI DNA barcodes for farm-associated mosquitoes and assess this data set for the presence of hidden diversity. Of the three DNA extraction methods assessed, the Hotshot technique was most rapid and we obtained good quality barcode sequences. In total, we analyzed 22 morphospecies in: *Aedes* (9 species), *Anopheles* (5 species), *Coquillettidia* (1 species), *Culex* (3 species), and *Culiseta* (4 species). Neighbor-joining analysis demonstrated that most specimens clustered to species recognized by morphology. Intraspecific sequence divergences within distinct species ranged from 0% to 2.07%, while higher interspecific divergences were found between pairs such as *Coquillettidia richardii/An. atroparvus* (19.71%) and *Culiseta morsitans/An. plumbeus* (19.86%). In the *Anopheles maculipennis* complex, discrepancy in certain specimens identified as *An. daciae* and *An. messeae* indicated the poor resolution of the COI DNA barcoding in separating these taxa. In the *Culex pipiens* complex, all specimens grouped within *Cx. pipiens pipiens*, and were well separated from *Cx. torrentium* and *Cx. modestus.* The existence of well-defined groups within *Aedes vexans* (Sweden and the Netherlands) and *Aedes cantans* (UK) indicate the likely inclusion of genetic diversity. In conclusion, DNA barcoding combined with a sound morphotaxonomic framework provided an effective approach for the identification of epidemiologically important mosquito's species in UK farm-associated mosquito species.

Fossil records and evolution of haematophagy in biting midges (Diptera: Ceratopogonidae)

R. Szadziewski, P. Dominiak, E. Sontag, A. Urbanek and J. Szwedo
University of Gdańsk, Department of Invertebrate Zoology and Parasitology, Wita Stwosza 59, 80-308 Gdańsk, Poland; heliocopris@gmail.com

We are not going to discuss which theory explains more accurately why dinosaurs went extinct, and what the main factors responsible for the evolutionary success of mammals were. Nor have we anything to say about the diseases affecting vertebrates from the Jurassic to the Neogene, even though some data on fossil parasitic protozoans can be found in the literature. However, we do intend to shed some light on the evolution of biting midges, a family of nematocerous flies containing species familiar as annoying bloodsucking pests and vectors of various pathogens, bothering vertebrate hosts probably ever since the Jurassic (over 176 Mya). Ceratopogonids, represented in the recent fauna by over 6,200 species, are well documented as fossils, especially in amber inclusions. More than 280 fossil species belonging to 48 genera (25 extant, 23 extinct) have been described, with the oldest one dated to 142 Ma (Lower Cretaceous). 112 of these species – from the subfamilies Lebanoculicoidinae and Leptoconopinae, the genera *Archiculicoides* and *Culicoides*, and the subgenus *Lasiohelea* of *Forcipomyia* – fed on vertebrate blood. Haematophagy is obviously a plesiomorphic condition in biting midges, and all basal lineages are considered to be bloodsucking parasites. The relict genera *Leptoconops* and *Austroconops*, as well as 6 extinct genera, are known from Lebanese amber (130 Ma). The genus *Culicoides* is younger, and to date has been reported from the French amber of Vendée (95-85 Ma) and New Jersey amber (93 Ma), however, our present study extends the history of this group back to 100 Ma. Other feeding habits among biting midges evolved later, mainly in the Upper Cretaceous (haemolymphophagous, predators) and the Paleogene (pollinophagous, nectarophagous, further predatory taxa). At the same time (45 Mya) haematophagy developed for the second time in the subgenus *Lasiohelea*. The evolution of feeding behaviours is correlated with morphological modifications, especially of the mouthparts, and the distribution of antennal sensilla coeloconica involved in host-seeking.

MALDI-TOF protein profiling for species identification of sand flies – practical lessons from the lab and from the field

V. Dvorak[1], K. Hlavackova[1], P. Halada[2], B. Alten[3], V. Ivovic[4] and P. Volf[1]
[1]Faculty of Science, Charles University in Prague, Department of Parasitology, Vinicna 7, 128444 Prague 2, Czech Republic, [2]Institute of Microbiology, Academy of Science, Videnska 1083, 142 00 Prague 4, Czech Republic, [3]Hacettepe University, Department of Biology, ESRL Laboratories Beytepe, 06800 Ankara, Turkey, [4]University of Primorska, Department of Biodiversity, Titov trg 4, 6000 Koper, Slovenia; icejumper@seznam.cz

Bloodfeeding females of phlebotomine sand-flies are the only proven vectors of *Leishmania* parasites, causative agents of leishmaniases, important human emerging diseases. Conclusive identification of vector species is profound. Protein profiling by matrix-assisted laser desorption/ionization time-of-flight (MALDI-TOF) mass spectrometry prooved to be a useful tool for species identification as it is simple, accurate, time- and cost-effective and requires minimal sample preparation. Methods of specimen capture, storage and sample preparation were optimized to obtain reproducible profiles and usage of different body parts was tested to develop a multi-approach protocol that utilizes a single specimen for identification by morphological, DNA- and protein-based methods. In sand fly females, the influence of blood meal and egg development was tested at several time intervals after feeding. Moreover, identification of blood meal origin was tested on females experimentally fed on different vertebrate hosts at various degrees of blood meal digestion by comparing protein profiles of abdomen containing host blood with protein profiles obtained from host blood only. In larvae (L2 to L4) and pupae of different stage of development protein spectra were acquired and compared with profiles of adult sand flies. Impact of larval diet on the quality of the protein profile was also tested. A reference spectra database was established and used for species identification of sand flies in two large-scale field surveys carried on in Balkan countries and Caucasian countries, respectively. All dubious samples where morphological and protein-based identifications did not match were clarified by a sequencing analysis, which confirmed the identification according to MALDI-TOF MS in all cases.

Genetic diversity of *Ixodes ricinus* and first evidence of *Ixodes inopinatus* ticks in Baltic countries

A. Paulauskas[1], E. Galdikaite-Braziene[1], J. Radzijevskaja[1] and A. Estrada-Peńa[2]
[1]Vytautas Magnus University, Vileikos 8, 44404 Kaunas, Lithuania, [2]University of Zaragoza, Calle de Pedro Cerbuna, 12, 50009 Zaragoza, Spain; a.paulauskas@gmf.vdu.lt

Ticks of Ixodidae family are wide spread parasitic arthropods that are involved in transmission of number of tick-borne pathogens. Baltic countries are located in a unique area of co-distribution of *Ixodes ricinus* and *I. persulcatus* ticks. In the present study, genetic variability of *I. ricinus* ticks collected in allopatric and sympatric locations in the Baltic countries has been investigated using a sequence analysis of the mitochondrial DNA control region, *16S rRNA* and cytb genes. There were 32 haplotypes (Hd: 0.8551) and 27 haplotypes (Hd: 0.8213) of control region sequences from ticks in allopatric and sympatric zones detected respectively. Out of 47 *16S rRNA* gene haplotypes, 32 haplotypes (Hd: 0.7213) were found in the allopatric zone and 27 (Hd: 0.9572) in the sympatric zone. *Cytb* gene was very conserved and monomorphic in ticks from the allopatric zone, whereas 3 unique haplotypes were observed in the sympatric zone. The higher number of unique haplotypes of control region was detected in the allopatric zone. Median joining network and Fst analysis did not reveal a clear separation between ticks from the two zones. Based on morphological and sequence analysis of *16S rRNA* gene 7 specimens were identified as *I. inopinatus*. Phylogenetic analysis shows clear separation from other *Ixodes* species. *I. inopinatus* (Estrada-Pena, Nava and Petney, 2014) is recently described species of Ixodidae family that originates in dry areas of the Mediterranean region in Spain, Portugal, Morocco, Algeria and Tunisia. It has also been collected in the areas of western Germany in sympatry with *I. ricinus*, far of its known distribution range. This is the first evidence of *I. inopinatus* ticks in Baltic countries.

Is the West Nile virus mosquito vector *Culex univittatus* present in the Iberian Peninsula? A comparative morphological and molecular analysis

V. Mixão[1], D. Bravo-Barriga[2], R. Parreira[1], M. Novo[1], C. Sousa[1], E. Frontera[2], L. Braack[3] and A.P.G. Almeida[1,3]
[1]Global Health and Tropical Medicine, Instituto de Higiene e Medicina Tropical, Universidade NOVA de Lisboa, R Junqueira 100, 1349-008 Lisboa, Portugal, [2]Parasitology and Parasitic Diseases, Veterinary Faculty, University of Extremadura, Av de la Universidad, CP10071 Cáceres, Spain, [3]Centre for Viral Zoonoses, Department of Medical Virology, University of Pretoria, Private Bag X20, Hatfield 0028, South Africa; palmeida@ihmt.unl.pt

Culex univittatus and *Culex perexiguus* mosquitoes (Diptera, Culicidae) are competent arbovirus vectors, but not always with clear morphological differentiation. In Europe, and in the Iberian Peninsula in particular, the presence of either or both species is controversial. However, in order to conduct adequate surveillance for arboviruses in this region, it is crucial to clarify whether *Cx. univittatus* is present or not, as well as to critically access existing differentiation tools. This study aimed to clarify this situation, by morphological and molecular comparison of Iberian specimens deemed as *Cx. univittatus*, with others of South African origin, i.e. from the type locality region. Midfemur pale line, hindfemur R ratio, setae g R1 ratio, seta f shape, length of ventral arm of phalosome and number of setae on IX tergal abdominal segment were observed. A phylogenetic analysis based on COI mtDNA, of which there were no sequences from *Cx. univittatus* available in GenBank, was performed. Iberian and South African specimens are morphologically similar, except for the length of the ventral arm of the phalosome, which was higher in the Iberian specimens. Although the Iberian specimens could not be accurately identified using BOLD Systems, phylogenetic analysis still grouped them closer to South African *Cx. univittatus*, than to *Cx. perexiguus* from Turkey. However, both taxa segregated as two individual monophyletic clusters with shared common ancestry. These results suggest that the Iberian mosquito's specimens are likely the West Nile virus vector *Cx. univittatus* as previously found.

Mosquito surveillance around Syrian and European borders of Turkey; DNA barcoding and virus screening

Y. Sarikaya[1], K. Ergunay[2], F. Gunay[1], S. Kar[3], K. Oter[4], S. Orsten[2], O. Kasap[1], Y. Linton[5] and B. Alten[1]
[1]Hacettepe University, Department of Biology, ESRL, 06800, Beytepe, Ankara, Turkey, [2]Hacettepe University, Faculty of Medicine, Department of Medical Microbiology, Virology Unit, 06100 Sihhiye Ankara, Turkey, [3]Namik Kemal University, Department of Biology, 59100 Tekirdag, Turkey, [4]Istanbul University, Faculty of Veterinary Medicine, Department of Parasitology, 34320 Istanbul, Turkey, [5]The Walter Reed Biosystematics Unit, MRC-534, Smithsonian Institution, Suitland, MD 20746-2863, USA; yasemensarikaya@gmail.com

The internal unrest in Syria forced more than 4,000,000 citizens to migrate to neighboring countries. This situation increased the importance of the surveillance of mosquitoes and mosquito born diseases in the area. Within this scope, mosquito samples were collected in the South East and North West borders where refugees are either settled in great numbers or passing through in order to arrive to the European Countries. During the mosquito season, sampling was performed in 10 provinces where traps were set in over 200 stations. After morphological identification mosquitoes were separated for DNA barcoding by sequencing mitochondrial COI gene region and virus screening via generic nested and real-time PCR assays targeting flavi and alphaviruses. Results from over 13,000 samples comprised 14 species including *Aedes caspius*, *Ae. pulcritarsis*, *Anopheles claviger*, *An. maculipennis s.l.*, *An. sacharovi*, *An. superpictus*, *Coquillettidia richiardii*, *Culex perexiguus*, *Cx. pipiens s.l.*, *Cx. theileri*, *Cx. tritaeniorhynchus*, *Culiseta annulata*, *Cs. longiareolata* and *Uranotaenia unguiculata*. Samples molecularly identified as *Ae. caspius*, *An. maculipennis s.s.*, *Cx. pipiens s.s.* and *Cs. annulata* from Thrace Region, *An. sacharovi* from Mediterranean Region were found positive for chikungunya virus. Additionally WNV strain 1 clade 1a was detected in *Cx. pipiens s.s.* samples from Thrace Region. The presence of CHIK virus in the country was recorded for the first time and revealed the ongoing circulation of West Nile. DNA barcode reference database created with molecular identification results is helpful to estimate the primary vector species in the critical areas.

Diptera fauna associated with dog and cat carcasses and their possible role as vectors of several agents of animal and human diseases

S.P. Diz[1], C.S. Loução[1*], I.P. Fonseca[1], M. Santos[1] and M.T. Rebelo[2]*
[1]CIISA, Faculty of Veterinary Medicine, University of Lisbon, Av. Universidade Técnica, 1300-477 Lisbon, Portugal, [2]CESAM, Faculty of Sciences, University of Lisbon, Department of Animal Biology, Campo Grande, 1749-016 Lisbon, Portugal; silvia_diz@hotmail.com; [*]These authors contributed equally to this work

Diptera or 'true flies' are one of the most widespread insect orders in the world, with a great capability to disperse and adapt to a wide variety of conditions. Thus, it is no wonder that Diptera constitute such efficient means in the propagation of infectious agents. They pose an important role as mechanical vectors, carrying pathogens from a contaminated material, e.g. a carcass. In the presence of carrion, a great quantity of insects is drawn to that area, which greatly increases the chance of pathogen transmission to nearby animals and humans. The objective was to contribute for the study of the entomological succession in domestic animal corpses, whilst also highlighting some of these species' potential as vectors of infectious diseases agents. Modified Malaise traps were placed at the Faculty of Veterinary Medicine of the University of Lisbon from August to December 2014 and along the year of 2015 in order to capture insects attracted to dog and cat carrion. Captured specimens were taxonomically classified based on their morphology and those reported for the first time in Portugal had their DNA extracted and sequenced. So far, three new species of Diptera have been reported for the first time in Portugal: *Lucilia cuprina*, *Pollenia leclercqiana* and *Sarcophaga aratrix*. *L. cuprina* in particular is infamous as a myiasis agent in both animals and humans, being known to be a mechanical vector of pathogens and resistant to some insecticides. The present study contributed for a better knowledge on local insect fauna collected post mortem in dogs and cats that died from different causes and also for Diptera seasonal dynamics and their possible role as vectors of animal and human disease agents.

A novel multiplex PCR assay for the identification of six Eastern Mediterranean phlebotomine sand fly species

I.A. Giantsis[1], A. Chaskopoulou[1] and M.C. Bon[2]
[1]European Biological Control Laboratory, USDA-ARS, Tsimiski 43, 54623 Thessaloniki, Greece, [2]European Biological Control Laboratory, USDA-ARS, Campus International de Baillarguet, 34988 Montferrier-sur-Lez, France; igiant@afs.edu.gr

Sand flies (Diptera: Psychodidae, subfamily Phlebotominae) are haematophagous insects that are known to transmit several anthroponotic and zoonotic diseases. Reliable identification of sand flies at species level is crucial for their surveillance, the detection and spread of their pathogens and the implementation of targeted pest control strategies. In the current study we designed a novel, time saving, cost effective and easy-to-apply multiplex PCR assay, that avoids sequencing, for the identification of 6 Eastern Mediterranean sand fly species: *Phebotomus perfiliewi*, *P. simici*, *P. tobbi*, *P. papatasi*, *Sergentomyia dentata* and *S. minuta*. This methodology is based on species-specific single nucleotide polymorphisms of the nuclear *18S rRNA* gene, using one common and six diagnostic primers. Amplification products were easily and reliably separated in agarose gel yielding 1 single and clear band of diagnostic size for each species. Further, we verified its successful application on tissue samples immersed directly to the PCR mix, skipping DNA extraction. This direct multiplex PCR can be completed in less than 3 hours including all operating procedures, and costing no more than a simple PCR. The applicability of this methodology in the detection of hybrids is an additional considerable benefit.

Occurrence of *Anopheles maculipennis* group mosquitoes (Diptera: Culicidae) in Portugal: ecological determinants and importance as disease vectors

H.C. Osório[1], J. Gomes[2], L. Zé-Zé[1] and M.J. Alves[1]
[1]National Institute of Health Dr. Ricardo Jorge, Centre for Vectors and Infectious Diseases Research, Av.ª da Liberdade, 5, 2965-575 Águas de Moura, Portugal, [2]National Institute of Agriculture and Veterinary Research, Laboratory of Parasitology, Av. da República, Quinta do Marquês, 2780-157 Oeiras, Portugal; hugo.osorio@insa.min-saude.pt

The *Anopheles maculipennis* complex includes mosquitoes which have been considered vectors of malaria in Europe. Three members have been reported in Portugal: *Anopheles maculipennis* (*s.s.*), *An. melanoon* (syn. *subalpinus*), and *An. atroparvus*. To identify and study the distribution of these sibling species and the potential presence of others, a previously described polymerase chain reaction (PCR) assay using the nuclear ribosomal internal transcribed spacer 2 (ITS2) was used. A total of 1,033 specimens (221 immatures; 91 males; 721 females – 674 in pools of 1-50 specimens) collected from 138 locations in Portugal were molecularly analyzed for species identification. Of these, the blood meal of 12 blood-fed females was determined by sequencing cytochrome-b partial fragments with DNA from female's abdomen. Moreover, 674 specimens were screened for flaviviruses by RT-PCR and 721 were screened for the presence of parasitic roundworms (Onchocercidae) by PCR. Only the species *An. atroparvus* and *An. maculipennis* were detected in this study. The first was found to be widespread, occurring in all surveyed regions in Portugal, whereas *Anopheles maculipennis* was only found in the northern regions. Eight out of ten blood meals analyzed were from mammalian source, including human. Regarding pathogens screening, all mosquito pools were RT-PCR negative for flaviviruses. However, five pools representing 53 *An. atroparvus* females were PCR positive for *Dirofilaria immitis*. All the five corresponding locations of these pools are in the south of Portugal. This study aims to contribute to the current knowledge of the distribution and role as vector of the *Anopheles maculipennis* species in Portugal. The correct identification and detailed knowledge of the biology of the species in the *Anopheles maculipennis* group are of relevance given their historical role as vectors of malaria in Europe and recent findings of *Anopheles* females infected with *Dirofilaria* spp. in Portugal.

A survey of the mosquito fauna of the middle and eastern Blacksea Region of Turkey

M.M. Akiner[1], M. Ozturk[1], F.M. Simsek[2], C. Karacaoglu[3] and S.S. Caglar[3]
[1]RTE University, Biology Department, RTE University Faculty of Arts and Sciences Department of Biology Zoology Section, 53100 Rize, Turkey, [2]Aydın Adnan Menderes University, Biology Department, Aydın Adnan Menderes University Faculty of Arts and Sciences Department of Biology Ecology Section, Aydın, Turkey, [3]Hacettepe University, Biology Department, Hacettepe University Faculty of Science Department of Biology Ecology Section Beytepe Kampus, 06532 Ankara, Turkey; akiner.m@gmail.com

The region of Middle and Eastern Blacksea stated eastern border of Turkey with Georgia and Armenia. This area is an important for agriculture and tourism. Area including rice fields area middle part and hazelnut, tea and kiwi plantation areas for eastern part. A mosquito survey in 15 cities for 181 locations was conducted from june 2014 to june 2016. Twenty species of mosquitoes in four genera were collected. Area included nine species of *Culex*, one species of *Culiseta*, four species of *Anopheles*, six species of *Aedes* genus. More common species of the all locations were *Culex pipiens* complex and *Anopheles maculipennis* complex species. Experiments of the molecular study revealed that the *Culex* species are generally *Cx. pipiens pipiens* and area included *Cx. pipiens molestus*, *Culex pipiens pallens* and some areas included hybrid populations. *Anopheles maculipennis* complex species were found compeletely *Anopheles maculipennis maculipennis* after molecular analysis. Common larval habitats included rice fields, overflowing fountain troughs, water filled animal footprint, drainage channels in middle parts and rainwater pond, drainage channels, water filled animal foodprint, inside used tire. Aduls were collected from animal shelters, inside houses and landing on humans during larval collection times. *Aedes aegypti* and *Aedes albopictus* firstly detected of established populations in the eastern part of Blacksea and Turkey. This study was funded by The Scientific and Technological Research Council of Turkey (project no. 113Z795) and partially funded by RTE University Scientific Resesearch Project Coordination Office (project no. 2012.102.03.08)

Sterilization of *Aedes japonicus japonicus* (*Hulecoeteomyia japonica japonica*) by high-energy photon irradiation: implications for a sterile insect technique approach in Europe

F. Balestrino[1], A. Mathis[1], S. Lang[2] and E. Veronesi[1]
[1]National Centre for Vector Entomology, Institute of Parasitology, Faculty of Veterinary Science (Vetsuisse), University of Zurich, Zurich, Switzerland, [2]Radiation Oncology Clinic, University Hospital, Zurich, Switzerland; f.balestrino@iaea.org

Aedes japonicus japonicus (*Hulecoeteomyia japonica japonica*) (Diptera: Culicidae), is attracting particular attention for its high spread distribution in Europe and its possible roles in the transmission of human and animal pathogens. The preferential habitats of this species are forested and bushy areas and the control and surveillance strategies have been reported extremely difficult and ineffective. The possible use of the sterile insect technique (SIT) against this species may substantially contribute to the integrated control of *Ae. j. japonicus* which favours ecological niches hardly accessible. The present study investigates the effects of irradiation at a dose of 40 Gy on fitness parameters in *Ae. j. japonicus*. Irradiation was performed using a TrueBeam linear accelerator without the involvement of nuclear sources. Long colonized males (Pennsylvania, PA, USA) were crossed with females of the same colony or with field-collected females (Zurich, Switzerland). Rates of blood feeding and fertility were lower when PA strain males were kept with field-collected females rather than PA females. Irradiated males induced reductions in fertility (residual fertility: 2.6%) and fecundity in mated females while seems not affecting the mating competitiveness of the males. The data indicate that the SIT is a suitable technique but the role of colonization methods need to be further investigated to fully understand the potentiality of SIT against *Ae. j. japonicus*.

Aedes aegypti in California: novel control strategies in response to the recent invasion

F.S. Mulligan and J. Holeman
Consolidated Mosquito Abatement District, P.O. Box 278, Selma, CA 93662, USA; smulligan@mosquitobuzz.net

The invasion of *Aedes aegypti* L. into dispersed areas of California, beginning in 2013, has created significant public health issues. A primary vector for dengue and chikungunya viruses, as well as zika virus, this mosquito is also closely associated with human habitation for oviposition sites and exhibits a preference for humans as hosts. In invaded neighborhoods, *Ae. aegypti* is a huge biting nuisance and source of resident complaints, yet it is difficult to control utilizing conventional treatment methods. Thus innovative approaches are called for in combating this important vector. During 2016, the Consolidated Mosquito Abatement District collaborated with MosquitoMate, Inc., to evaluate their novel sterile insect technique against a recently established *Ae. aegypti* population within a small neighborhood of Clovis, CA, USA. This SIT incorporates the mass rearing and release of large numbers of males infected with the bacterium *Wolbachia* in order to mate with non-infected wild females. Such matings result in cytoplasmic incompatibility and egg infertility. We will discuss the elements and issues involved in development of the study; including the selection of study sites, gaining access and acceptance from homeowners, development of procedures and protocols specific to the efficacy evaluation of this SIT method against *Ae. aegypti* in an arid habitat, as well as dealing with media and providing and disseminating public education materials.

The BG-Counter, the first operative automatic mosquito counting device for online mosquito monitoring: field tests and technical outlook

A. Rose[1], M. Weber[1], I. Potamitis[2], P. Villalonga[3], C. Pruszynski[4], M. Doyle[4], M. Geismar[5], J. Encarnação[3] and M. Geier[1]
[1]Biogents AG, Weißenburgstr., Regensburg, Germany, [2]Technological Educational Institute of Crete, E Daskalaki, Rethymno, Greece, [3]Irideon S.L., C. St. Isidre, Alaior, Spain, [4]Florida Keys Mosquito Control District, 107th Street Gulf, Marathon, FL 33050, USA, [5]Universität Regensburg, Zoologie, Regensburg, Germany; andreas.rose@biogents.com

As part of an autonomous trap station, the BG-Counter automatically differentiates captured mosquitoes from other insects, counts them, and wirelessly transmits results to a cloud server for further analysis. It can also collect and transmit data like temperature, rH, precipitation, air movements, or light intensity. Thus, it allows for a real-time monitoring of local mosquito populations, as well as variables that influence their development and activity. The collected data can be accessed, displayed and analysed on a cloud-based dash board, accessible on the internet. They can also be exported to Excel. Thus, vector control professionals can now follow the mosquito situation with an unprecedented accuracy, while overcoming constraints associated with manual inspection. Adulticiding can be performed when mosquitoes are most active, the effectiveness of control measures can be validated immediately. The collected data can give insights into variables that influence mosquito activity, supporting research into the development of more efficient and environmentally friendly mosquito control techniques. In most mosquito control situations, the target species and their biology is known, and the identification of the captured mosquito species is unnecessary. In situations like the surveillance of invasive mosquitoes or the monitoring of vectors during epidemics, species identification would however be important. In our presentation, we describe the technical background of the BG-Counter and supply examples of data sets generated in field. We also present initial results from our research into the advanced version of the BG-Counter, which is planned to be able to identify and differentiate species of special interest. Supported by the EU's 7th Framework Programme (grant 306105, acronym MCD) and the EU's Horizon 2020 programme (grant 691131, acronym REMOSIS).

Expression of anti-*Trypanosoma* sp. proteins in *Sodalis glossinidius* in the model *Glossina morsitans morsitans* (Diptera: Glossinidae)

S. Tudela Zúquete[1], S. Van Harten[2], A.P. Basto[1], M.O. Afonso[3] and L. Alfaro Cardoso[1]
[1]Faculdade de Medicina Veterinária, Universidade de Lisboa, Centro de Investigação Interdisciplinar em Sanidade Animal (CIISA), Avenida da Universidade Técnica, 1300-477 Lisboa, Portugal, [2]Faculty of Veterinary Medicine, Lusofona University of Humanities and Technologies, CBIOS – Research Center for Bioscience and Health Technologies, Campo Grande 376, 1740-024 Lisboa, Portugal, [3]Instituto de Higiene e Medicina Tropical (IHMT), Medical Parasitology Unit, Global Health and Tropical Medicine (GHTM), Rua da Junqueira, 100, 1349-008 Lisboa, Portugal; sarazuquete@fmv.ulisboa.pt

Tsetse flies (*Glossina* spp.) are responsible for the transmission of the flagellated protozoa *Trypanosoma* sspp. which causes animal African trypanosomiasis (Nagana) and Human African trypanosomiasis. The later is endemic in 30 countries in sub-Saharan Africa and it is estimated that 60 million people may be at risk of infection. Climate and environmental changes are likely to increase the incidence of the human pathology as well as its geographical distribution. Strategies undertaken to fight African trypanosomiasis will have to be multidisciplinary and articulated between the different components that comprise its biological system. Actions must be directed, as well as applied, together with the populations. Public governmental entities should be involved in a good governance exercise. The development of molecular biology techniques has opened up new possibilities with respect to vector control. Despite the fact that the direct transgenesis of flies is hampered by tsetse´s adenotrofic viviparity, paratransgenesis technique emerged as an alternative, presenting itself more advantageous, nowadays. In order to develop an endosymbiont of *Glossina* spp. able to express recombinant proteins with trypanocidal effect (attacin and defensin), different techniques were applied for the transformation of *Sodalis glossinidius*. These techniques included thermal shock, chemical treatment and electroporation. Transformation of *S. glossinidius* by a combination of 2 of these tecniques was, for the best of our knowledge, successfully achieved first time. The endosymbiont was transformed with different plasmid constructs to express attacin and defensin proteins, individually.

The *Anopheles gambiae* 1000 Genomes Project: a resource for vector control research

C.S. Clarkson[1], A. Miles[2], N.J. Harding[2], G. Bottà[3], D.P. Kwiatkowski[1,2] and A. gambiae 1000 Genomes Project[1,2]
[1]Wellcome Trust Sanger Institute, Hinxton, Wellcome Trust Genome Campus, United Kingdom, [2]Oxford University, Oxford, Wellcome Trust Centre for Human Genomics, United Kingdom, [3]University of Rome, La Sapienza, Rome, Italy; cc28@sanger.ac.uk

The *Anopheles gambiae* 1000 Genomes Project (Ag1000G) is conducting whole-genome deep sequencing of wild-caught malaria vectors from populations across Africa with three core objectives: discover genetic variation, describe population structure/history and to connect these with malaria epidemiology and vector ecology. In phase 1 of the project, sequence data from 765 specimens from eight countries have been generated. The data have been used to discover over 52 million SNPs, on average one SNP every two accessible bases, providing the first genome-wide view of the spectacular diversity within natural populations. These data have been publicly released and constitute the largest open access genomic resource available for any vector species. Here we provide an overview of the Ag1000G phase 1 data resource and initial results of population genetic analyses, focusing on applications to vector control. Analyses of population structure reveal a complex mosaic, with incomplete speciation, geography, demography and selection all influencing gene flow across the species' range. We illustrate this with the *Vgsc* gene, the target site for DDT and pyrethroid insecticides, where we show multiple independent haplotypes were involved in selective sweeps, shared between species and between populations separated by thousands of kilometres, indicating an extraordinary potential for mutations to spread. We also present preliminary evidence for novel resistance mutations, discussing how haplotype data from Ag1000G could be used to track the spread of resistance. Finally, we present data from the Ag1000G coastal Kenyan population, where individuals exhibit long runs of homozygosity consistent with a severe recent population bottleneck. These data demonstrate that demographic events leave a strong genetic signal, and we discuss how genomic data could be used to provide feedback about the impact of vector control interventions.

Suppression of vectors of dengue, chikungunya and zika diseases using super-sterile *Aedes* males

P. Kittayapong[1], S. Ninphanomchai[1], S. Khaklang[1], W. Limohpasmanee[2], U. Chansang[3], C. Chansang[3] and P. Mongkalangoon[4]
[1]Mahidol University, Centre of Excellence for Vectors and Vector-Borne Diseases, Faculty of Science, Salaya, Nakhon Pathom 73170, Thailand, [2]Thailand Institute of Nuclear Technology, Ongkharak, Nakhon Nayok 26120, Thailand, [3]Ministry of Public Health, Department of Medical Sciences, Tivanond Road, Nonthaburi 11000, Thailand, [4]Ministry of Public Health, Department of Diseases Control, Tivanond Road, Nonthaburi 11000, Thailand; pkittayapong@gmail.com

Arboviral diseases such as dengue, chikungunya, and zika diseases are considered as global important public health problems. These diseases are transmitted mainly by *Aedes aegypti*, a domestic day-time biting mosquitoes, which distributed through out tropics and sub-tropics. So far implemented vector control programs that could have sustainable and significant impacts on these mosquito vectors have not yet been demonstrated. The main objective of our research is to develop an alternative environmental friendly approach that could suppress mosquito vector populations and eventually reduce disease incidences. Recently, development of the super-sterile *Aedes* vectors has been achieved by combining *Wolbachia*-induced incompatibility approach with sterile insect technique (SIT) using radiation. *Wolbachia*-transinfected *Ae. aegypti* lines have been developed by direct microinjection using double *Wolbachia* strains from *Ae. albopictus*. Laboratory experiments showed no difference in competitiveness of irradiated males when compared to normal ones as well as complete sterility of normal females after mating with super-sterile males. A proof-of-concept in suppressing natural populations of *Ae. aegypti* mosquito vectors using this strategy is currently on-going in Chachoengsao Province, eastern Thailand. One-year data on household abundance and egg hatching rate of targeted mosquito vectors were used to form the baseline for monitoring the release of super-sterile males. Community engagement and public awareness through media have resulted in positive support for practical use of this strategy in wider areas.

The use of Sentinel 1 SAR imagery for wide area mosquito control

S. Mourelatos, S. Gewehr, S. Kalaitzopoulou, C.G. Karydas, M. Iatrou and G. Iatrou
Ecodevelopment S.A., Thesi Mesaria, P.O. Box 2420, 57010 Filyro, Greece; mourelatos@ecodev.gr

The goal of this study is the evaluation of Sentinel-1 Synthetic Aperture Radar imagery of the European Space Agency (S-1 SAR, ESA) for operational purposes in wide area mosquito control projects trough monitoring of surface water dynamics and mosquito population data. The study area is the plain of Thessaloniki, Northern Greece and the investigation period is spring/summer 2016. Sentinel-1A launched in 2014, in combination with Sentinel-1B, launched in April 2016, provide an all-weather, day-and-night radar imagery of Earth's surface every six days. With an analysis of 5 m/pixel, S-1 SAR imagery is considered to be ideal for rapid change detection analysis at a scale of a 25 m^2 unit surface. Performance of S-1 SAR images to detect surface water will be assessed with ground information from field data of the mosquito monitoring networks of Ecodevelopment, namely a larvae monitoring network (>20,000 sampling stations throughout the region) and an adult monitoring network (80 fixed CO_2 light traps throughout the region and 10 human landing sampling stations within the rice fields). Special emphasis will be given to: (a) the flooding event of 20th-21st of May 2016 in Central Macedonia following intensive rainfalls (~80 mm in 36 hrs); and (b) the inundation scheme in the rice fields of Thessaloniki plain (~22,000 ha) from May to July 2016. The current study will present methods for modelling adult mosquito populations, based on S-1 SAR imagery data, larvae and adult monitoring data and the spatial distribution of the flooding events. The models will be developed based on support vector machines (SVMs), partial least squares (PLS), principal component analysis (PCA) and the multilayer perceptron (MLP). The performance of these models will be evaluated. The ground truthing of the results of the analysis of S-1 SAR imagery is expected to be extremely valuable for the detection and updating of the breeding sites within and beyond the existing network of larvae sampling stations of Ecodevelopment. In addition, the results of the change detection analysis of surface waters will be evaluated for a possible correlation with mosquito adult population dynamics.

Response of *Aedes albopictus* (Skuse) males to acoustic and visual stimuli

F. Balestrino[1], D.P. Iyaloo[2], K.B. Elahee[2], A. Bheecarry[2], F. Campedelli[3], M. Carrieri[3] and R. Bellini[3]
[1]International Atomic Energy Agency, FAO/IAEA Division of Nuclear Techniques in Food and Agriculture, Wagramerstrasse 5, 1400 Vienna, Austria, [2]Ministry of Health & Quality of Life, Vector Biology & Control Division, SSR Botanical Garden Rd., Curepipe, Mauritius, [3]Centro Agricoltura Ambiente, Medical and Veterinary Entomology, Via Argini Nord 3351, 40014 Crevalcore, Italy; f.balestrino@iaea.org

Aedes albopictus (Skuse) is one of the most invasive mosquito species capable of transmitting various harmful pathogens to humans. Failure of vector control strategies against this species requires the development of new effective vector control methods. Among the alternative genetic control measures under development, the sterile insect technique (SIT) is today receiving a renovated interest as a possible effective tool to be integrated in an area-wide pest management approach. The monitoring of the abundance, distribution, movements and ratio of released sterile and wild fertile males is a fundamental requirement for the successful management of any pest control activities integrating an SIT component. Although several traps exist for adult female mosquito, effective traps to monitor large quantities of males were less investigated and difficult to obtain. In this study we analyzed the response of *Ae. albopictus* males to various sound stimuli produced with different volumes and frequencies in association with visual cues for the optimization of male catches. The production of modulated frequencies continuously varying within the typical female sound emission range (500-650 Hz) showed the best results for a sound pressure level between 75 and 79 dB (at the speaker level). The black color of the trap cylinder, however, seems decisive to attract males in the vicinity of sound traps. We also confirmed that males increase their response to acoustic stimulation up to 4 days of age and then show a continuous and progressive decline of their sound responsiveness. A sound trap prototype producing the most effective sound stimuli proven at laboratory conditions was also successfully developed and tested in the field in Mauritius. The use of sound stimuli therefore appears a promising prospect to increase the capture rate of *Ae. albopictus* males in dedicated or in already existing mosquito traps.

Development of Etofenprox-treated clothing for the US Military

U.R. Bernier
USDA-ARS-CMAVE, MFRU, 1600 SW 23rd Drive, Gainesville, FL 32606, USA; uli.bernier@ars.usda.gov

Historically, casualties from diseases have greatly outnumbered those from combat during military operations. Since 1951, US military combat uniforms have been chemically treated to protect personnel from arthropod attack. In the 1970s and 1980s, permethrin was one of several insecticides evaluated as a repellent treatment for uniforms. In 1991, permethrin became the standard treatment of US military combat uniforms. In 2007 the US Marine Corps transitioned from treatment with permethrin in the field to factory treatment of their 50/50 nylon/cotton Marine Corps Combat Utility Uniforms (MCCUUs). The US Army transitioned to factory treatment of uniforms in 2009. Over the past few years, an increasing proportion of combat uniforms are constructed from fabric comprised of nylon, rayon and fire resistant materials such as para-aramid or meta-aramid. These uniforms cannot be treated with in the field and must therefore be treated at the factory level. In 2015, the evaluation of etofenprox-treated clothing was completed and the data were submitted to US EPA for registration. Etofenprox has an improved toxicological profile compared to permethrin and due to this, it can be impregnated into clothing at a higher concentration. The study was conducted using Fire-Resistant Army Combat Uniform type III fabric and evaluated with fabrics that were washed 75 times. The treated uniforms were highly protective out to 75 wash cycles. Results from bite protection studies will be covered in this presentation.

The passive gravid *Aedes* trap (BG-GAT): an environmentally-friendly, low-cost and effective tool for *Aedes* (*Stegomyia*) surveillance

J.K. Mccaw[1], B.J. Johnson[2], A.E. Eiras[3] and S.A. Ritchie[2]
[1]Biogents AG, Weissenburgstrasse 22, 93047 Regensburg, Germany, [2]James Cook University, Australian Institute of Tropical Health and Medicine, P.O. Box 6811, Cairns, Queensland 4870, Australia, [3]Universidade Federal de Minas Gerais, Departamento de Parasitologia, LabEQ, Belo Horizonte, Minas Gerais, Brazil; jennifer.mccaw@biogents.com

The current circulation of zika, on top of the various other viruses that are transmitted by *Aedes aegypti* and *Aedes albopictus*, has once again shown the importance of sensitive and effective surveillance tools for these invasive species. The recently developed passive Gravid *Aedes* Trap (BG-GAT) is an effective, practical, low cost, and easily transportable tool to collect gravid *Aedes (Stegomyia)* species. The BG-GAT is essentially a lethal oviposition trap where trapped mosquitoes are killed, usually by contact insecticides, and can be easily collected and identified during trap inspections. Such features are essential in large-scale monitoring programs. In semifield observations, the BG-GAT captured a significantly higher proportion of gravid *Ae. aegypti* than the double sticky ovitrap. Further field trials in Cairns and the USA compared the BG-GAT to existing sticky ovitraps and the BG-Sentinel trap. In Cairns, the potential for the BG-GAT to be used for dengue virus surveillance was also demonstrated. While field trials in Brazil demonstrated low capture rates due to a high resistance to pyrethroids, a new publication investigated several alternative insecticide and insecticide-free killing agents for use in the BG-GAT. These included long-lasting insecticide-impregnated nets (LLINs), vapor-active synthetic pyrethroids (metofluthrin), canola oil, and two types of dry adhesive sticky cards. These results demonstrated that the use of inexpensive and widely available insecticide-free agents could be an effective alternative to pyrethroids in regions with insecticide-resistant populations. The BG-GAT can be a useful tool for capturing adult *Ae. aegypti* and *Ae. albopictus* and may be suitable for other container-breeding species such as *Culex quinquefasciatus*. The low cost, practicality of operation and the high catch rates make the BG-GAT suitable for vector surveillance and projects requiring monitoring of mosquitoes for arboviruses.

Evaluation of a *Wolbachia* transinfected *Aedes albopictus* strain as a vector population suppression agent

A. Puggioli[1,2], M. Calvitti[3], R. Moretti[3] and R. Bellini[2]
[1]Università di Bologna, Dipartimento di Scienze e Tecnologie Agroambientali, Viale G. Fanin 42, 40127 Bologna, Italy, [2]Centro Agricoltura Ambiente G. Nicoli, Medical and Veterinary Entomology Department, Via Argini Nord 3351, 40014 Crevalcore, Italy, [3]ENEA, Biotechnologies and Agroindustry Division, Via Anguillarese, 301, 00123 Rome, Italy; apuggioli@caa.it

AR *w* P *Aedes albopictus* line was developed thanks to the artificial infection with a heterologous *Wolbachia* strain, resulting in a bidirectional incompatibility pattern with wild-type *Ae. albopictus.* AR *w* P was tested for its suitability to support intensive rearing conditions as required for mass production and field release. In this study, we compared AR *w* P and wild-type *Ae. albopictus* strains, both reared under standard operating procedures, evaluating the following functional parameters: (1) male and female pupation dynamics; (2) efficiency in mechanical sexing; and (3) male mating competitiveness in comparison with irradiated and wild-type males. AR *w* P males showed a higher production rate of male pupae in the 24 hours after pupation onset and a lower percentage of residual contaminant females when applying mechanical sexing procedures. AR *w* P males were more efficient than wild-types in competing for wild-type females in large enclosures, thus inducing a level of sterility significantly higher than that expected for an equal mating competitiveness. These results encourage the use of this *Ae. albopictus* strain as suppression tool against *Ae. albopictus.*

Field evaluation of AQUATAIN AMF™ as a mosquito larval and pupal control agent in different breeding sites in Northern Italy

R. Veronesi[1], *M. Carrieri*[1], *A. Albieri*[1], *S. Di Cesare*[1], *M. Panziera*[1], *G. Bazzocchi*[2] *and R. Bellini*[1]
[1]*Centro Agricoltura Ambiente G. Nicoli, Medical and Veterinary Entomology, Via Argini Nord 3351, 40014 Crevalcore (BO), Italy,* [2]*Bleu Line Srl, Via Virgilio, 28 Zona Industriale Villanova, 47122 Forlì, Italy; rveronesi@caa.it*

Due to the EU policy developed in the last years in the frame of the Biocide directive, mosquito control sector is facing a serious problem due to the shortage in insecticide products availability. The restriction in the spectrum of insecticides may also increase the risk of resistance in target population. New products with different action mechanisms are therefore required to strengthen the tools box currently available in Europe. AQUATAIN AMF™ is a silicone based monomolecular surface film with a physical mode of action blocking the larvae and pupae breathing, which is considered a good agent for the management of resistance to commonly used larvicides. During the summer 2016 we conducted several field trials in order to evaluate the efficacy and the lasting activity of AQUATAIN AMF™ in the most important mosquito breeding sites such as rice fields, irrigation ditches and road drains in Northern Italy. The target mosquito species were *Aedes caspius* and *Culex pipiens* in the rural settings (rice fields and ditches) and *Aedes albopictus* and *Culex pipiens* in the urban settings (road drains). Results will be presented and discussed in the poster.

Comparative evaluation of six outdoor sampling traps for disease-transmitting mosquitoes in reference to human landing catch in rural Tanzania

A.J. Limwagu, E.W. Kaindoa, N.S. Matowo, L.F. Finda, S.A. Mapua, A.S. Mmbando and F.O. Okumu
Ifakara Health Institute, Environmemntal Health and Ecological Science-Thermatic Group, Off Mlabani Passage, P.O. Box 53, Tanzania; alimwagu@ihi.or.tz

There is a growing concern on how mosquito sampling methods can be safely performed in malaria endemic countries. Human landing catch is the best mosquito sampling method, it is labor intensive and exposes individual to malaria transmission risks. The ongoing study is assessing the different traps and aiming to find an alternative for HLC in terms of effectiveness, densities, diversities and behaviors of disease-transmitting mosquitoes. About seven traps, Mosquito Magnet (MMX), BG-Sentinel, Suna trap, Ifakara Tent Trap-C (ITT-C), M-Trap and M-Trap fitted with CDC Light trap were used. The traps were comparatively evaluated and calibrated with reference to the human landing catch (HLC). 7×7 latin square experiments were conducted in 6 different villages in 12 months, working in dry and wet season in each of the villages. Seven position identified to each of this villages with the distance of 100 m from one trap to another. The different traps rotated to the seven positions, that at the end of a 7 day rotation, each trap type had been to each of the seven locations at least once. The experiment were replicated 3 times for 21 nights, start from 18:00 hrs to 06:00 hrs. The outcome measure will be the comparisons of effectiveness of different traps in terms of capturing high density and diversities of outdoor host seeking mosquitoes relatively to the reference method (HLC). A total of 62,317 of all female mosquitos were collected for six villages for both seasons wet and dry, where BG-Sentinel n=5,571 (8.94%), HLC n=13,909 (22.32%) ITT-C n=3,775 (6.06) MMX n=4,468 (7.17%) M-Trap n=8,429 (13.52) M-Trap with CDC Light trap n=12,011 (19.27%) and Suna trap n=14,003 (22.47%). The result is showing there is no significant different between HLC, Suna trap and M-trap fitted with CDC for all total number of female mosquitoes but there is a different between HLC, against M-trap, BG-Sentinel, MMX and ITT-C in the first round.

S-methoprene insect growth regulator mosquito larvicides result of efficacy trials

D.B. Bajomi[1], J.S. Schmidt[1], L.T. Takacs[1] and B.S. Serrano[2]
[1]Babolna Bio Ltd., Szallas utca 6., 1107 Budapest, Hungary, [2]T.E.C. Laboratory, 1, rue Jules Vedrines, 64600 Anglet, France; bajomi.daniel@babolna-bio.com

Global warming, mass tourism and delivery of goods helps the appearance and proliferation not only of new vector species, but very serious illnesses too. During the past years the rapid Southern-Northern spread of the invasive tiger mosquito (*Aedes albopictus*) has been well noted. The zika-virus battle has just started. While the safety of humans remains the major target, the available tools are rapidly decreasing, as a number of insecticides and larvicides are either lost, or their use is more and more restricted because of the European Biocidal Product Regulation. In some of the EU Member States chemical mosquito treatment is still the most common, while the more efficient and environmentally safer biological larviciding remains less favoured. S-methoprene insect growth regulator based formulations have been successfully used for decades, especially in the USA, Canada, Australia, but less in Europe. A number of formulations has been recently developed and introduced primarily against *Aedes* and *Culex* species. The S-methoprene based BIOPREN® granule, liquid, tablet and pump spray formulations permit their use in a wide range of water systems and different environmental conditions. According to the laboratory and field studies, 2 to 6 weeks' residuality can be achieved. Some formulations are suitable for professional, others for amateur use. From resistance management point of view the interchange use of Bti and S-methoprene formulations are of extreme importance. The poster presentation wishes to introduce these formulations, target species, their application and efficacy results.

Using conical Biogents suction traps to reduce populations of *Aedes albopictus* and *Aedes aegypti* in a residential area in Houston, TX

M. Reyna[1], M. Debboun[1], J. Vela[1], C. Roberts[1], O. Salazar[1], S. Gordon[2] and M. Geier[2]
[1]Harris County Public Health and Environmental Services, Mosquito Control Division, 3330 Old Spanish Trail, Houston, TX 77021, USA, [2]Biogents AG, Weißenburgstr. 22, 93055 Regensburg, Germany; scott.gordon@biogents.com

We used a prototype trap that incorporates the same ventilator and perforated cover as the BG-Sentinel, but instead of the collapsible bucket, has a conical textile trap body which requires suspension. Similar to the BG-Sentinel trap, it attracts vectors through a combination of visual cues and an artificial lure that mimics human skin odors. The goal of the study was to determine if a network of conical traps in a defined residential neighborhood could achieve a local reduction in container breeding *Aedes* in Houston, TX, USA. A treatment cluster, consisting of 5 adjoining houses, was established where 10 traps were placed and run continuously as the intervention tool. A control cluster consisting of 5 adjoining houses with no traps was also established. The primary outcome of the study was to detect the effect of BG conical mosquito traps on nuisance biting by *Ae. albopictus* and *Ae. aegypti*, as measured by human landing counts (HLC). The secondary outcome was to determine the abundance of *Ae. albopictus* and *Ae. aegypti* based on BG Sentinel monitoring collections. In the treatment site, median numbers of *Ae. albopictus* in HLCs were significantly lower than in the control site, resulting in an 81% reduction of *Ae. albopictus.* Although there were no significant differences in the median numbers of *Ae. albopictus* females collected in weekly BG-Sentinel monitoring traps between the two sites, about 33% fewer *Ae. albopictus* were collected in the treatment BG-Sentinels. Regarding *Ae. aegypti*, significantly more females were collected in the control site compared to the treatment site yielding a reduction of more than 70%. The use of attractant baited traps for selective removal trapping of *Aedes* mosquitoes offers an attractive alternative to the conventional approach of using chemical insecticides. Traps are free of pesticides and do not contribute to the development of resistance. Catching host-seeking females on a daily basis will reduce the biting pressure and the risk of disease transmission.

Mosquito Mist Nets for the surveillance of disease vectors

B. Domènech[1,2], E. Molins[1], A. Rose[3] and K. Paaijmans[2,4]
[1]Institut de Ciència de Materials de Barcelona, Campus UAB, 08193 Bellaterra, Spain, [2]Barcelona Institute for Global Health, C/ Rosselló, 132, 08036 Barcelona, Spain, [3]Biogents AG, Weißenburgstr. 22, 93055 Regensburg, Germany, [4]Centro de Investigação em Saúde de Manhiça, Rua 12, Vila de Manhiça, Mozambique; berta.domenech@isglobal.org

To-date, most successful mosquito traps collect a specific part of the vector population; mosquitoes that are looking for a blood-meal, a resting site, or a water body to oviposit. While these tools have been proven very valuable and provide great insight into mosquito ecology and behaviour, they often tend to collect only a specific subset of a mosquito population. Our aim is to develop a technology to collect random mosquito samples, regardless of species, sex, age, feeding status, or host or oviposition preferences, thus providing more representative and realistic information on the overall population structure and infection status at a given location and point in time. The idea was dubbed Mosquito Mist Net (MMN) and consists of a net with an electrostatic charge and a sticky surface, having the ability to trap mosquitoes and to prevent them from escaping or from falling down, easing their collection for further research. Initial tests of nets with an electrostatic charge (our initial idea of improving mosquito captures) and different electrode materials (e.g. iron, aluminium, polymeric insulating screens) showed that mosquitoes could sense the electric field and actively avoided it. Therefore the focus was shifted to adhesives alone. In order to coat such netting and ease mosquito collection, a new water-soluble coating adhesive for the net based on Polyvinyl-alcohol and Polyvinylpyrrolidone was developed. This adhesive remains sticky for several days and can be easily washed with water, easing their use on the field. Moreover, first mosquito behavioural tests results showed a mosquito attachment, comparable to commercial permanent sticky adhesive materials.

Evaluation of lavender, eucalyptus and orange essential oils as alternative repellents against the *Ixodes ricinus* ticks

M. Kulma, T. Bubová and F. Rettich
The National Institute of Public Health, Šrobárova 48, 100 42, Praha 10, Czech Republic; kulma@af.czu.cz

The essential oils are considered to be a user and environment friendly alternative of synthetic repellents. Therefore, in this study we evaluated repellent effects of three (lavender, eucalyptus and orange) essential oils, with no allergen or mutation data reported, and of DEET against adult females of *Ixodes ricinus*. Performed bioassay showed that repellency of all investigated non-synthetic oils decreased over time, while the effects of DEET remained relatively stable during the whole experiment. Initial repellency (after 15 minutes) of investigated samples was found to be moderate to high: 65% (orange) to 85% (lavender). At the end of the investigation (after 90 minutes), lavender oil still showed the highest repellency effect of the tested oils (45%). At this point, eucalyptus oil repelled only 15% of ticks and orange essential oil did not show any repelling effect against ticks since the 20 minute mark. In contrast, the efficiency of DEET ranged between 95-100% throughout the whole experiment. In conclusion, this study revealed, that tested oils act as repellents, but they could not be considered to be equal to or as stable as the synthetic DEET. Further investigation into the technology process optimization, which could lead to increasing the effectivity of essential oils, is thus needed. On the other hand, essential oils, especially lavender, showed interesting potential to become an alternative repellent for shorter outdoor activities.

Benefits of combining transfluthrin-treated sisal decorations with LLINs against indoor and outdoor biting malaria vectors

J.P. Masalu, F.O. Okumu and S.B. Ogoma
Ifakara Health Institute, Environment Health and Ecological Sciences Thematic Group, P.O. Box 53, Ifakara, Morogoro, Tanzania; jpaliga@ihi.or.tz

Transfluthrin vapour disrupts mosquitoes' host seeking and prevents bites. The goal of this study was to measure benefits of combining transfluthrin treated sisal decorations with long lasting insecticidal nets (LLINs) against indoor and outdoor biting malaria vectors. The protective efficacy of transfluthrin treated sisal baskets and hessian flags were measured in terms of reduced indoor density, exposure to bites and mortality of malaria vectors in experimental huts and against outdoor biting vectors in outdoor restaurants using human landing catches. Sisal decorative baskets treated with 2.5 ml and 5.0 ml transfluthrin deterred three quarters (relative rate (RR) [95% CI] = 0.26 [0.2, 0.34; $P<0.001$, RR [95% CI] = 0.29 [0.22, 0.37]; $P<0.001$) of *Anopheles arabiensis* mosquitoes from entering huts. Furthermore, transfluthrin treated items increased the probability of mortality by 2 fold (OR [95% CI] = 2.69 [2.09, 3.47]; $P<0.001$ and OR [95% CI] = 3.45 [2.71, 4.40]; $P<0.001$) and reduced exposure to outdoor biting *An. arabiensis* mosquitoes by more than 90% (RR [95% CI] = 0.11 [0.09, 0.15]; $P<0.001$ and RR [95% CI] = 0.14 [0.11, 0.18]; $P<0.001$). This study demonstrates that locally produced transfluthrin treated sisal decorative baskets and flags provide additional protection against early evening indoor and outdoor bites of malaria vectors when people are not using LLINs. Interestingly, transfluthrin treated items increased mortality of mosquitoes that were caught inside huts even in the presence of LLINs.

Chemosensory responses to the repellent *Nepeta* oil by the yellow fever mosquito, a vector of zika virus

J.T. Sparks[1,2], J.D. Bohbot[2,3], M. Ristić[4], D. Mišić[5], M. Skorić[5], A. Mattoo[6] and J.C. Dickens[2]
[1]Department of Biology, High Point University, High Point, NC, USA, [2]United States Department of Agriculture, Agricultural Research Service, Henry A. Wallace Beltsville Agricultural Research Center, Invasive In, Beltsville, MD, USA, [3]Department of Entomology, Hebrew University, Rehovot, Israel, [4]Institute for Medicinal Plants Research 'Dr Josif Pančić', University of Belgrade, Tadeuša Košćuška 1, 11000, Belgrade, Serbia, [5]Institute for Biological Research 'Siniša Stanković', University of Belgrade, Bulevar despota Stefana 142, 11060, Belgrade, Serbia, [6]United States Department of Agriculture, Agricultural Research Service, Henry A. Wallace Beltsville Agricultural Research Center, Sustainable, Beltsville, MD, USA; joseph.dickens@ars.usda.gov

The insect repellent properties of catnip (*Nepeta* essential oil) and its major component, nepetalactone, have long been recognized. However, the neural mechanisms through which these repellents are sensed by mosquitoes, including the yellow fever mosquito *Aedes aegypti*, an important vector of zika virus, were poorly understood. Here we show that volatiles of these repellents activate olfactory receptor cells within basiconic sensilla on the maxillary palps of female yellow fever mosquitoes. A taste receptor cell on the labella of females was sensitive to both the feeding deterrent quinine and the repellents. Activity of a second taste receptor cell sensitive to the feeding stimulant sucrose was suppressed by both repellents. Our results provide chemosensory pathways for the reported spatial repellency and feeding deterrence properties of these repellents. Knowledge of the sensory input utilized by female mosquitoes when choosing to feed will facilitate design of new management strategies involving the use of repellents.

Knockdown resistance of *Culex pipiens* complex species populations in the Middle and Eastern Blacksea regions of Turkey

M. Ozturk[1], M.M. Akiner[1], F.M. Simsek[2] and C. Karacaoglu[3]
[1]RTE University, Biology Department, RTE University Faculty of Arts and Sciences Department of Biology Zoology Section, 53100 Rize, Turkey, [2]Aydın Adnan Menderes University, Biology Department, Aydın Adnan Menderes University Faculty of Arts and Sciences Department of Biology Ecology Section, Aydın, Turkey, [3]Hacettepe University, Biology Department, Hacettepe University Faculty of Science Department of Biology Ecology Section Beytepe Campus, 06532 Ankara, Turkey; muratoztrk29@gmail.com

Insecticide resistance in mosquito populations is a main obstacle for control of both mosquitoes and mosquito-borne diseases. Knockdown resistance (kdr) is one of the mechanisms of resistance against pyrethroids. *Culex pipiens* complex species are main WNV vectors and spreading urban and suburban areas in Turkey. In this study, we aimed to investigate kdr mutation of *Culex pipiens* complex species in middle and eastern Blacksea Turkey. Larvae and egg rafts collected by larval dipping whole suitable areas and transferred to the laboratory. After emergence of adults were identified using morphological identification keys and molecularly confirmed which complex species. For this purpose, genomic DNA isolation was performed from different populations. Extracted DNA was used for molecular identification and kdr mutation screening. Samples identified after molecularly visualisation and separate which complex species for *kdr* screening. Second PCR prosess was performed for kdr and the amplicons sent to sequencing for the detect kdr mutations after viaulisation with electrophoresis. As a result of the scanning mutation point from 30 collection points, three diffrent mutation types were found. One mutation is a silence and another mutations lead to leucine-phenylalanine changing. 16 collection points are heterozygote (leucine-phenylalanine) resistant, 4 collection points are homozygote (phenylalanine-phenylalanine) resistant, 8 collection points are homozygote (leucine-leucine) susceptible, 2 collection points have including silence mutation and heterozygote susceptible phenotype. Resistance frequency of the area (heterozygote and homozygote) is 66.66% and important for the control operations. This study was funded by The Scientific and Technological Research Council of Turkey (project no. 113Z795)

Mapping and forecasting mosquito problems using freely accessible remote sensing data

M. Schäfer
Biological Mosquito Control/NEDAB, Vårdsätravägen 5, 75644 Uppsala, Sweden; martina.schafer@mygg.se

Mosquito appearance is generally connected to water, with the species having their defined environmental preferences. In Sweden and other countries, mosquito problems are caused by floodwater mosquito species that require flat areas flooded via adjacent rivers or lakes, or by precipitation alone. Thus, data on slope, vegetation, soil and precipitation could be used to predict occurrence of potential breeding sites. This data used to be difficult to find and expensive, but recently satellite-based data has become a valuable and affordable data source. In particular, the start of the Copernicus-program with the Sentinel-satellites of the European Space Agency is remarkable with almost all data freely accessible. This includes radar data and multispectral data which both are extremely valuable for remote sensing of potentially flooded areas. Some examples on how this data can be used will be shown. In an earlier study in the Dalälven region in Central Sweden, we have shown that a time series of radar data from both dry and wet periods can be used to identify temporary flooded areas, by performing PCA and unsupervised classification. With freely available data, the utilization and effectivity of satellite-based data offers many possibilities.

Effect of mass rearing conditions on mosquito life history traits and on mating competitiveness of sterile male *Anopheles arabiensis* in semi field

D.D. Soma[1,2], H. Maïga[2,3], W. Mamai[3], N.S. Bimbilé-Somda[2,3], C. Venter[4], R.S. Lees[5], F. Fournet[1,6], A. Diabaté[2], K.R. Dabiré[2] and L.R.J. Gilles[3]
[1]Master International d'Entomologie, MIE, Université de Montpellier/CEMV, France, [2]Institut de Recherche en Sciences de la Santé/Centre Muraz, BP 545, Bobo Dioulasso, Burkina Faso, [3]International Atomic Energy Agency, IPCL, Vienna, Austria, [4]Wits Research Institute for Malaria, FHS, Johannesburg, South Africa, [5]London School of Hygiene and Tropical Medicine, Keppel Street, London, United Kingdom, [6]Institut de Recherche pour le Développement, IRD, MIVEGEC, France; dieusoma@yahoo.fr

Since the early stages of the sterile insect technique, it has been recognized that the mating competitiveness of sterile insects is a critical factor for its successful application. The main objective of this study was to assess in semi field cages the effect of mass-rearing on mating competitiveness of sterile male *Anopheles arabiensis. An. arabiensis* immature stages were reared in large scale using larval tray-rack unit and on a small scale using small laboratory rearing trays. Several life history traits such as pupation rate, emergence rate, body size as well as the effect of irradiation on adult longevity were evaluated. Moreover, mating competitiveness between sterile mass-reared males and fertile males reared in small scale competing for small scale virgin females was also investigated. Five to six day-old mosquitoes left for two nights of mating with a ratio of 1:1:1 (100 irradiated males: 100 non-irradiated males: 100 virgin females) compared to controls of 0:100:100 (non-irradiated control) and 100:0:100 (irradiated control) were used. Competitiveness was determined by assessing the egg hatch rate from five replicates. No significant difference in most of the life history parameters between rearing methods was observed. Regardless the rearing method, the competitiveness index of sterile males compared to controls was 0.58, indicating that sterile males were half as competitive as non-irradiated males. Overall, our results showed that the mass rearing process did not affect mosquito life history parameters and the mating competitiveness of mass reared males.

Attraction of mosquitoes and stable flies to toxic baits designed for house fly management

J.A. Hogsette and D.L. Kline
USDA-ARS-CMAVE, Mosquito and Fly Research Unit, 1600 S. W. 23rd Drive, Gainesville, FL 32608, USA; jerry.hogsette@ars.usda.gov

Attractants to lure mosquitoes and stable flies to traps or toxic baits are usually different than those used for house flies. Many mosquito traps depend on CO_2 or other host odors; stable fly traps reflect light in attractive wavelengths. House flies are attracted to sugars and proteinaceous materials, e.g. egg yolks. Two companies wanted their house fly toxic baits evaluated against stable flies and mosquitoes. The first bait was Zyrox Fly Granular Bait (cyantraniliprole 0.5% AI, Syngenta Crop Protection, Greensboro, NC, USA). For the evaluation, 25 stable flies, 3-5 days old, were placed in small (11×11×18 cm long) cages, all with 10% sucrose solutions. Three control cages treated with just 10% sucrose were paired with three cages treated with 15 ml of Zyrox and 10% sucrose. Evaluations began about 10:30 am. Mortality was counted about 8:30 am and 2:30 pm daily. Forty eight hours after the evaluation began, mean mortality was 93.3% in the treated cages and 0.0% in the controls. The Zyrox bait AI must be ingested to produce mortality. These blood-feeding flies had a choice between the sucrose and the bait and they consumed enough bait to produce morality. The second bait was Yellow End Zone Fly Stickers (EZFS) (acetamiprid 4.4% AI, FMC Corp., Philadelphia, PA, USA). Fly stickers are a flexible plastic with the AI applied to the surface in a sugar solution. Stickers (11×11 cm) were placed vertically in the end of the cages described above. Six treated cages with stickers and 10% sucrose solution were paired with six control cages containing only 10% sucrose. Each cage had 25, 3-5 day old stable flies. Evaluations began in the morning, and mortality was counted 6 hr later and at 24-h intervals thereafter. Mean mortality was 93.3% after a 30-h exposure. There are two novel behavioral observations associated with this test. Stable flies will rest on vertical surfaces, but there is no reason to expect them to rest on the yellow EZFS more than any other cage surface. Color may be a factor. Also if they did rest on the EZFS, there is little reason to expect that feeding would occur. Obviously it did. Results were similar for adult *Ae. aegypti* mosquitoes.

Evaluation of pyriproxyfen against *Aedes aegypti*: small-scale field trials in Madeira Island-Portugal

B. Pires[1], G. Seixas[1], G. Alves[1], A. Jesus[1], A.C. Silva[2], R. Paul[3], G. Devine[4] and C.A. Sousa[1]
[1]GHTM-Instituto de Higiene e Medicina Tropical-UNL, Rua da Junqueira 100, 1349-008 Lisboa, Portugal, [2]Departamento de Saúde, Planeamento e Administração Geral, IASAUDE, IP-RAM, Rua das Pretas 1, 9004-515 Funchal, Portugal, [3]Functional Genetics and Infectious Disesases Unit, Institut Pasteur, 25-28 Rue du Docteur-Roux, 75724 Paris, France, [4]QIMR Berghofer Medical Research Institute, 300 Herston Rd, Herston-Queensland 4006, Australia; casousa@ihmt.unl.pt

Pyriproxyfen (PPF), an insect juvenile hormone mimetic, is a promising compound for the control of *Aedes aegypti* (Linnaeus, 1762). This larvicide interferes with the mosquito development by inhibiting pupal emergence and can be carried, from treated to untreated locations, on mosquito legs. *Aedes aegypti* was identified during 2005 in Funchal, Madeira. This highly insecticide resistant population expanded throughout the Southern coast of the island and, thus, effective vector control tools are necessary to reduce mosquito population. Our study aimed to evaluate the effectiveness of PPF auto-dissemination by *Ae. aegypti* in Paúl do Mar and Funchal municipalities, as a possible tool to be used in control strategies. The study was divided in 3 phases: (a) pre-treatment: assessment of mosquito densities and immature mortality in the absence of PPF; (b) treatment: PPF dissemination and mortality rates estimates in artificial breeding sites; and (c) post-treatment assessment of mosquito densities. Mosquito densities were assessed with BG-traps and ovitraps. Adapted BG-traps were also used as dissemination stations. The PPF dissemination was evaluated based on larval and pupal mortality of laboratory-reared specimens, placed in artificial breeding sites (ABS) deployed throughout the intervention areas. Results showed PPF auto-dissemination in both study areas affecting the development and survival of ABS larvae. Immature mortality occurred mainly in ABS close to the dissemination stations confirming the limited dispersal of this species. Significant decrease in egg densities also occurred in both intervention areas, after PPF dissemination. Based on these results, PPF may be a promising tool for the control of *Ae. aegypti* in Madeira. Funding: SFRH/BD/98873/2013; GHTM – UID/Multi/04413/2013; FP7-HEALTH-DENFREE (282378).

Protective efficacy of transfluthrin repellent treated footwear against mosquito bites: semi-field system and full-field setting evaluation

T. Gavana, P. Sangoro and P.P. Chaki
Ifakara Health Institute, Environmental Health and Ecological Sciences, Mikocheni A, Kiko Avenue, P.O. Box 78373, 255 Dar es Salaam, Tanzania; tgavana@ihi.or.tz

Despite dramatic reduction of malaria over the past two decades, the wide use of insecticide treated nets (ITNs) and indoor residual spraying (IRS) will not be sufficient to eliminate the disease in endemic countries especially in areas where mosquito vectors bite predominantly in the early evenings and rest outdoors and people spend their evenings outdoors. Novel efficacious and effective spatial repellents delivery means may be a useful tool to complement ITNs and IRS. This study evaluated the protective efficacy of repellent impregnated footwear both in the semi-field system and full-field setting. We treated strips of locally made sandals with 8% concentration of a spatial repellent. The treated and untreated sandals were then worn by the volunteers and tested against malaria vectors in the semi-field system with insectary reared *Anopheles arabiensis* and in the full field with wild mosquitoes. Each experiment was conducted for 24 consecutive nights by two groups of male volunteers, one group wearing repellent treated sandals (intervention) and another group wearing sandals without repellent (control). In the semi-field system, treated sandals provided 68.87% protection [RR=0.31 (95% C.I. = 0.25-0.38, $P<0.0001$, z=-11.29)] against *Anopheles arabiensis* bites when compared to control. In the full-field setting, the sandals provided 70.80% personal protection against the wild *Anopheles gambiae s.l.* [RR=0.22 (95% C.I. = 0.14-0.37, $P\leq0.0001$, z=-5.93]. Repellent impregnated footwear demonstrated to confer significant protection against malaria mosquito bites in the pseudo-field and field settings. This design of repellent delivery presents a potential tool to supplement the currently used interventions for mosquito control.

List of participants

First name	Last name	Email address	Affiliation	Country
Mustafa	Akiner	akiner.m@gmail.com	RTE University	Turkey
Alessandro	Albieri	aalbieri@caa.it	Sustenia S.r.l.	Bosnia
Ana Margarida	Alho	alhinha@gmail.com	Faculdade de Medicina Veterinária, Universidade de Lisboa	Portugal
Paulo	Almeida	palmeida@ihmt.unl.pt	IHMT/UNL/PT	Portugal
Bulent	Alten	kaynas@hacettepe.edu.tr	Hacettepe University	Turkey
Daniel	Bajomi	bajomi.daniel@babolna-bio.com	Bábolna Bio Ltd.	Hungary
Mame Tierno	Bakhoum	thierno.bakhoum@cirad.fr	CIRAD	France
Fabrizio	Balestrino	f.balestrino@iaea.org	Insect Pest Control Laboratory, Joint FAO-IAEA Division of Nuclear Techniques in Food and Agriculture, International Atomic Energy Agency	Austria
Carlos	Barceló	carlos.barcelo@uib.es	UIB	Spain
Frederic	Bartumeus	fbartu@ceab.csic.es	CEAB-CSIC & CREAF	Spain
Stavroula	Belere	smpeleri@esdy.edu.gr	National School of Public Health, Dep. of Parasitology, Entomology and Tropical Diseases	Greece
Christelle	Bender	cbender@demoustication-bas-rhin.fr	SLM67	France
Amal	Bennouna	bennouna1amal@gmail.com	Hassan II Agronomy and Veterinary Institute	Maroc
Dirk	Berkvens	dberkvens@itg.be	Institute of Tropical Medicine	Belgium
Ulrich	Bernier	uli.bernier@ars.usda.gov	USDA-ARS-CMAVE	USA
Elena	Bogdanova	nekton-zieger@mail.ru	I.M. Sechenov First Moscow State Medical University	Russia
Marieta	Braks	marieta.braks@rivm.nl	National Insitute of Public Health and the Environment	the Netherlands
Tamiko	Brown-Joseph	tamiko.brown-joseph@sta.uwi.edu	School of Veterinary Medicine, Faculty of Medical Sciences,The University of the West Indies	Republic of Trinidad and Tobago
Katharina	Brugger	katharina.brugger@vetmeduni.ac.at	Vetmeduni Vienna	Austria
Rubén	Bueno	rbueno@lokimica.es	Laboratorios Lokímica	Spain
Carina	Carvalho	carina-l-carvalho@sapo.pt	Centre for vectors and infectious diseases (CEVDI), National Institute of Health Doutor Ricardo Jorge (INSARJ)	Portugal
Silvia	Ciocchetta	silvia.ciocchetta@qimrberghofer.edu.au	QIMR Berghofer Medical Research Institute	Australia

First name	Last name	Email address	Affiliation	Country
Chris	Clarkson	cc28@sanger.ac.uk	Wellcome Trust Sanger Institute	UK
Vito	Colella	vito.colella@uniba.it	University of Bari	Italy
Lorna	Culverwell	lorna.culverwell@helsinki.fi	Department of Virology – Haartman InstituteUniversity of Helsinki	Finland
Jetske	De Boer	jetske.deboer@wur.nl	Wageningen University	the Netherlands
Ine	De Goeyse	idegoeyse@itg.be	Institute for Tropical Medicine Antwerp	Belgium
Rita	De Sousa	rsr.desousa@gmail.com	National Institute of Health Dr. Ricardo Jorge	Portugal
Berna	Demirci	demirciberna80@gmail.com	Kafkas University	Turkey
Major	Dhillon	mdhillon@northwestmvcd.org	SOVE	USA
Joseph	Dickens	oseph.dickens@ars.usda.gov	USDA, ARS, BARC, IIBBL	USA
Marian	Dik	m.t.a.dik@nvwa.nl	Food and Consumer Product Safety Authority (NVWA), Centre for Monitoring of Vectors	the Netherlands
Silvia	Diz	silvia_diz@hotmail.com	Faculdade de Medicina Veterinária	Portugal
Ivana	Djuri	ivanaveterinar@gmail.com	Institute for Biocides and Medical Ecology	Serbia
Berta	Domènech	berta.domenech@isglobal.org	Institut de Salut Global de Barcelona (ISGlobal)	Spain
Patrycja	Dominiak	heliocopris@gmail.com	Department of Invertebrate Zoology and Parasitology, University of Gdansk	Poland
Vit	Dvorak	icejumper@seznam.cz	Dept. of Parasitology, Faculty of Science, Charles University in Prague	Czech Republic
Roger	Eritja	reritja@elbaixllobregat.cat	Consell Comarcal del Baix Llobregat	Spain
Thahsin	Farjana	thahsinfarjana@gmail.com	Bangladesh Agricultural University	Bangladesh
Ina	Ferstl	inschen-f@web.de	University of Freiburg	Germany
Eleonora	Flacio	eleonora.flacio@supsi.ch	LMA-SUPSI	Switzerland
Claire	Garros	claire.garros@cirad.fr	Cirad, UMR Controle des Maladies Animales Exotiques et Emergentes/CYROI	France
Sandra	Gewehr	gewehr@ecodev.gr	Ecodevelopment S.A.	Greece
Ioannis	Giantsis	igiant@afs.edu.gr	USDA-ARS	Greece

First name	Last name	Email address	Affiliation	Country
Maria	Goffredo	m.goffredo@izs.it	Istituto Zooprofilattico Sperimentale dell'Abruzzo e del Molise	Italy
Scott	Gordon	scott.gordon@biogents.com	Biogents AG	Germany
Ulla	Gordon	ulla.gordon@biogents.com	Biogents AG	Germany
Filiz	Gunai	gunayf@gmail.com	Hacettepe University	Turkey
Nausicaa	Habchi-Hanriot	nhabchihanriot@terra.com	Institut Pasteur de la Guyane	French Guiana
Hella	Heidtmann	hella.heidtmann@tiho-hannover.de	Institute for Parasitology, University of Veterinary	Germany
Luis Miguel	Hernandez-Triana		Animal and Plant Health Agency	UK
Jaime	Herrezuelo Antolín	jaime.herrezuelo@gmail.com	Instituto Cavanilles (Universitat de València)	Spain
Eva C	Heym	eva.heym@zalf.de	Leibniz Centre for Agricultural Landscape Research (ZALF)	Germany
Kristyna	Hlavackova		Department of Parasitology, Faculty of Science, Charles University in Prague	Czech Republic
Adnan	Hodzic	adnan.hodzic@vetmeduni.ac.at	Institute of Parasitology, University of Veterinary Medicine Vienna	Austria
Jerome	Hogsette	jerry.hogsette@ars.usda.gov	USDA-ARS-CMAVE	USA
Adolfo	Ibanez-Justicia	a.ibanezjusticia@nvwa.nl	Centre for Monitoring of Vectors of the Netherlands	the Netherlands
Aleksandra	Ignjatovic Cupina	cupinas@polj.uns.ac.rs	University of Novi Sad, Faculty of Agriculture	Serbia
Famke	Jansen	fjansen@itg.be	Institute of Tropical Medicin	Belgium
Ana	Jesus	ana.spj20@gmail.com	Instituto de Higiene e Medicina Tropical, Universidade Nova de Lisboa	Portugal
Ricardo	Jimenez Peydró	jimenezp@uv.es	Universitat de València	Spain
Artur	Jöst	artur.joest@kabs-gfs.de	Kommunale Aktionsgemeinschaft zur Bekämpfung der Schnakenplage (KABS)	Germany
Frederic	Jourdain	frederic.jourdain@ird.fr	IRD	France
Katja	Kalan	katja.kalan@upr.si	Faculty of Mathematics, Natural Sciences and Information Technologies University of Primorska	Slovenia
Abdoulaye	Kane Dia		Laboratoire d'Ecologie Vectorielle et Parasitaire	Dakar
Maria	Kazimirova	maria.kazimirova@savba.sk	Institute of Zoology, Slovak Academy of Sciences	Slovakia

First name	Last name	Email address	Affiliation	Country
Heli	Kirik	hekirik@gmail.com	Estonian University of Life Sciences	Estonia
Martin	Kulma	martin.kulma@szu.cz	National Institute of Public Health	Czech Republic
Gulcan	Kuyucuklu	gulcankuyucuklu@gmail.com	Medical Faculty	Turkey
Gregory	Lanzaro	gclanzaro@ucdavis.edu	University of California, Davis	USA
Tereza	Lestinova	terka.kratochvilova@seznam.cz	Charles University in Prague	Czech Republic
Danica	Liebenberg	danica.liebenbergweyers@nwu.ac.za	North-West University	South Africa
Isabel	Lopes de Carvalho	isabel.carvalho@insa.min-saude.pt	INSA	Portugal
David	López Peña	david.lopez@uv.es	Universitat de Valencia	Spain
Javier	Lucientes	jlucien@unizar.es	University of Zaragoza	Spain
Peter	Luethy	peter.luethy@micro.biol.ethz.ch	Institute of Microbiology, ETH, Zurich	Switzerland
Renke	Lühken		Bernhard Nocht Institute for Tropical Medicine	Germany
Jan	Lundström	jan.lundstrom@mygg.se	Swedish Biological Mosquito Control/NEDAB	Sweden
Giovanni	Marini	giovanni.marini@fmach.it	University of Trento	Italy
Mihai	Marinov	mihai.marinov@ddni.ro	Danube Delta National Institute for Research and Development	Romania
Bruno	Mathieu	bmathieu@unistra.fr	Institute of Parasitology and Tropical Pathology	France
Jennifer	McCaw	jennifer.mccaw@biogents.com	Bioagents AG	Germany
Jason	Migdal	jmigdal@alumni.rvc.ac.uk	Royal Veterinary College	UK
Ognyan	Mikov	mikov@ncipd.org	NCIPD	Bulgaria
Miguel Angel	Miranda	ma.miranda@uib.es	University of the Balearic Islands	Spain
Tim	Möhlmann	tim.mohlmann@wur.nl	Linköping University/Wageningen University	the Netherlands
Tomás	Montalvo		Agencia Salud Pública de Barcelona	Spain
Spiros	Mourelatos	mourelatos@ecodev.gr	Ecodevelopment S.A.	Greece
Pie	Müller	pie.mueller@unibas.ch	Swiss Tropical and Public Health Institute	Switzerland
Gabi	Müller	gabi.mueller@zuerich.ch	Department of Health and Environment Zürich	Switzerland
Gabriela	Nicolescu	gabrielamarianicolescu@yahoo.co.uk	Cantacuzino' National Institute of Research, Medical Entomology	Romania

First name	Last name	Email address	Affiliation	Country
Kyohei	Nishino	nishinok@who.int	WHO Geneva	Switzerland
Ana	Norte	aclaudia.norte@gmail.com	CEVDI/ INSA & MARE	Portugal
Hugo	Osório	hugo.osorio@insa.min-saude.pt	National Institute of Health Dr. Ricardo Jorge	Portugal
Enrique	Padial Ramos	enrique.padial@icloud.com	North Bristol NHS Trust	UK
Young-Seuk	Park	parkys@khu.ac.kr	Kyung Hee University	Republic of Korea
Ilaria	Pascucci	i.pascucci@izs.it	Istituto zooprofilattico sperimentale dell'Abruzzo e del mMolise	Italy
Algimantas	Paulauskas	a.paulauskas@gmf.vdu.lt	Vytautas Magnus University	Lithuania
Isabel	Pereira da Fonseca	ifonseca@fmv.ulisboa.pt	Faculdade de Medicina Veterinária	Portugal
Yvon	Perrin	yvon.perrin@ird.fr	IRD	France
Branislav	Pesic	banekomarci@gmail.com	Institute for Biocides and Medical Ecology	Serbia
Dusan	Petric	dusanp@polj.uns.ac.rs	Laboratory for Medical and Veterinary Entomology, Faculty of Agriculture, University of Novi Sad	Serbia
Aleksandra	Petrovic	petra@polj.uns.ac.rs	Faculty of Agriculture, University of Novi Sad	Serbia
Françoise	Pfirsch	fpfirsch@demoustication-bas-rhin.fr	SLM67	France
Wolf Peter	Pfitzner	wolf-peter.pfitzner@kabs-gfs.de	KABS/IfD	Germany
Marie	Picard	marie.picard@ird.fr	IRD	France
Anita	Plenge-Boenig	anita.plenge-boenig@hu.hamburg.de	Institute for Hygiene and Environment, City of Hamburg	Germany
Olivier	Pompier	opompier@demoustication-bas-rhin.fr	SLM67	France
Arianna	Puggioli	aalbieri@caa.it	Centro Agricoltura Ambiente 'G. Nicoli'	Bosnia
Valeria	Purcarea-Ciulacu	airelav41@gmail.com	Cantacuzino National Institute of Research	Romania
Jana	Radrova	radrova@natur.cuni.cz	Faculty of Science, Charles University in Prague	Czech Republic
David	Ramilo	dwrramilo@hotmail.com	Faculdade de Medicina Veterinária	Portugal
Silvia	Ravagnan	sravagnan@izsvenezie.it	IZSVe	Italy
Frantisek	Rettich	frantisek.rettich@szu.cz	National Institute of Public Health	Czech Republic
Paula	Rodriguez		Instituto de Investigaciones Biológicas y Tecnológicas (IIBYT- CONICET/UNC)	Argentina

First name	Last name	Email address	Affiliation	Country
Andreas	Rose	andreas.rose@biogents.com	Biogents AG	Germany
Ignacio	Ruiz	irarrondo@riojasalud.es	CIBIR	Spain
Pedro	Sánchez-López	pedrof.sanchez2@carm.es	Murcia Regional Health Council (Spain)	Spain
Marcos	Santos	socramitis@msn.com	Faculdade de Medicina Veterinária	Portugal
Maria Margarida	Santos-Silva	m.santos.silva@insa.min-saude.pt	Instituto Nacional de Saúde Dr Ricardo Jorge	Portugal
Yasemen	Sarikaya	yasemensarikaya@gmail.com	Hacettepe University	Turkey
Martina	Schäfer	martina.schafer@mygg.se	Biological Mosquito Control/NEDAB	Sweden
Francis	Schaffner	fschaffner.consult@gmail.com	Francis Schaffner Consultancy	Switzerland
Sabine	Schicht	sabine.schicht@tiho-hannover.de	Institute for Parasitology, University of Veterinary	Germany
Marcus	Schmidt	marcus.schmidt@zuerich.ch	Umwelt- und Gesundheitsschutz Zürich	Switzerland
Gonçalo	Seixas	gseixas@ihmt.unl.pt	Instituto de Higiene e Medicina Tropical, Universidade Nova de Lisboa	Portugal
Magdalena	Sikora	magdalena.sikora9@gmail.com	Institute of Public Health for the Osijek-Baranya County	Croatia
Cornelia	Silaghi	cornelia.silaghi@uzh.ch	Institute of Parasitology	Switzerland
Fatih	Simsek	fsimsek@adu.edu.tr	Adnan Menderes University	Turkey
Charles	Smith	cwsmithskeeter@yahoo.com	Consolidated Mosquito Abatement District	USA
Charlotte	Sohier	charlotte.sohier@coda-cerva.be	CODA-CERVA (Veterinary and Agrochemical Research Centre)	Belgium
Seynabou	Sougoufara	seynabou.sougoufara@gmail.com	URMITE(Unité de Recherche sur les Maladies Infectieuses et Tropicales Emergentes)	Senegal
Carla	Sousa	casousa@ihmt.unl.pt	Instituto de Higiene e Medicina Tropical, Universidade Nova de Lisboa	Portugal
Jeroen	Spitzen	jeroen.spitzen@wur.nl	Wageningen University	the Netherlands
Arjan	Stroo		CMV-NVWA, Centre for Monitoring of Vectors, Netherlands Food and Product Safety Authority, Ministry of Economic Affairs	the Netherlands
David	Sullivan	zanusco1@msn.com	Zancor	USA

First name	Last name	Email address	Affiliation	Country
Milena	Svobodová	milena@natur.cuni.cz	Charles University in Prague	Czech Republic
Wesley	Tack	lvangorp@avia-gis.com	Avia-GIS	Belgium
Thomas	Van Loo	tvanloo@itg.be	Institute of Tropical Medicine	Belgium
Silvia	Vasconcelos	silvia_vasconcel@hotmail.com	Ed Por do Sol, Bl C 4º BJ,Fonte do Livramento, Caniço	Portugal
Laura	Vavassori	l.vavassori@unibas.ch	Swiss Tropical and Public Health Institute	Switzerland
Niels	Verhulst	niels.verhulst@wur.nl	Wageningen University	the Netherlands
Kenneth	Vernick	kvernick@pasteur.fr	Institut Pasteur	France
Rodolfo	Veronesi	rveronesi@caa.it	Centro Agricoltura Ambiente G. Nicoli	Italy
Eva	Veronesi	eva.veronesi@uzh.ch	National Centre for Vector Entomology, Institute of Parasitology, Zurich University	Switzerland
Veerle	Versteirt		Avia-GIS	Belgium
Chantal	Vogels	chantal.vogels@wur.nl	Laboratory of Entomology Wageningen University	the Netherlands
Melanie	Walter	melanie.walter@vetmeduni.ac.at	Vetmeduni Vienna	Austria
Doreen	Walther		ZALF	Germany
Lanjiao	Wang	wanglanjiao@gmail.com	Institut Pasteur of French Guiana	French Guiana
Graham	White	gbwhite@ufl.edu	University of Florida	USA
William	Wint	william.wint@zoo.ox.ac.uk	ERGO	UK
Nadja	Wipf	nadja.wipf@unibas.ch	Swiss Tropical and Public Health Institute	Switzerland
Renée	Zakhia	renee.zakhia@gmail.com	Lebanese University	Lebanon
Marija	Zgomba	mzgomba@polj.uns.ac.rs	Laboratory for Medical and Veterinary Entomology, Faculty of Agriculture, University of Novi Sad	Serbia
Carina	Zittra	carina.zittra@vetmeduni.ac.at	Institute of Parasitology	Austria
Almedina	Zuko	alma.zuko@vfs.unsa.ba	Veterinary Faculty of Sarajevo	Bosnia
Sara	Zúquete	sarazuquete@fmv.ulisboa.pt	Faculdade de Medicina Veterinária	Portugal

Author index

S

T

Printed in the United States
by Baker & Taylor Publisher Services